2nd International Conference on Automation, Control and Robotics Engineering (CACRE 2017)

AA006400

IOP Conference Series: Materials Science and Engineering
Volume 235

Prague, Czech Republic
28 – 30 June 2017

ISBN: 978-1-5108-4778-1
ISSN: 1757-8981

Printed from e-media with permission by:

Curran Associates, Inc.
57 Morehouse Lane
Red Hook, NY 12571

Some format issues inherent in the e-media version may also appear in this print version.

Printed by Curran Associates, Inc. (2017)

Institute of Physics
Dirac House, Temple Back
Bristol BS1 6BE UK

Phone: 44 1 17 929 7481
Fax: 44 1 17 920 0979

techtracking@iop.org

Additional copies of this publication are available from:

Curran Associates, Inc.
57 Morehouse Lane
Red Hook, NY 12571 USA
Phone: 845-758-0400
Fax: 845-758-2633
Email: curran@proceedings.com
Web: www.proceedings.com

Table of contents

Volume 235

2nd International Conference on Automation, Control and Robotics Engineering (CACRE 2017)

28–30 June 2017, Prague, Czech Republic

Accepted papers received: 1 September 2017
Published online: 8 September 2017

Control and Vision Engineering

IOP Conf. Series: Materials Science and Engineering **235** (2017) 011001 doi:10.1088/1757-899X/235/1/011001

PREFACE

CACRE 2017 was held in Prague, Czech Republic during June 28-30, 2017. CACRE 2017 was organized by Hong Kong Society of Mechanical Engineers(HKSME), supported by Czech Technical University in Prague and other organizations. The conference provides a useful and wide platform both for display the latest research and for exchange of research results and thoughts in broad Automation, Control and Robotics Engineering. The participants of the conference were from almost every part of the world, with background of either academia or industry, even well-known enterprise. The success and prosperity of the conference is reflected high level of the papers received.

The proceedings are a compilation of the accepted papers and represent an interesting outcome of the conference. This book cover 2 chapters: Robotics Engineering; Control and Vision Engineering.

We would like to acknowledge all of those who supported CACRE 2017. Each individual and institutional help were very important for the success of this conference. Especially we would like to thank the organizing committee for their valuable advices in the organization and helpful peer review of the papers.

We sincerely hope that CACRE 2017 will be a forum for excellent discussions that will put forward new ideas and promote collaborative researches. We are sure that the proceedings will serve as an important research source of references and the knowledge, which will lead to not only scientific and engineering progress but also other new products and processes.

Prof. Dan Zhang, York University, Canada
Prof. Xiaoping Liu, Carleton University, Canada

IOP Conf. Series: Materials Science and Engineering **235** (2017) 011002 doi:10.1088/1757-899X/235/1/011002

COMMITTEES

General Chair

Prof. Dan Zhang, York University, Canada

Conference Committee Co-Chairs

Prof. Xiaoping Liu, Carleton University, Canada

Prof. Du Ruxu, The Chinese University of Hong Kong, Hong Kong

Program Committee Co-Chairs

Prof. Tshilidzi Marwala, University of Pretoria, South Africa

Prof. Luděk Žalud, Brno Univeristy of Technology, Czech

Prof. Songyi Dian, Sichuan University, China

Local Chair

Dr. Martin Saska, Czech Technical University in Prague, Czech

International Technical Committees

Prof. Cho-Pei Jiang, National Formosa University, Taiwan

Prof. Shinn-Liang Chang, National Formosa University, Taiwan

Prof. Meng-Hui Hsu, Kun Shan University, Taiwan

Prof. Robert Babuska, Delft University of Technology, Netherlands

Prof. Filippo Arrichiello, University of Cassino and Southern Lazio, Italy

Prof. Nan-Chyuan Tsai, National Cheng Kung University, Taiwan

Prof. Genci Capi, Hosei University, Japan

Prof. Spilios Fassois, the University of Patras, Greece

Dr. Ali GhaffarianHoseini, AUT University, New Zealand

Dr. Darwin Lau, the Chinese University of Hong Kong

IOP Conf. Series: Materials Science and Engineering **235** (2017) 011002 doi:10.1088/1757-899X/235/1/011002

Dr. Giuseppe Carbone, University of Cassino and South Latium, Italy

Dr. Xiang Ping, City University of Hong Kong

Dr. Felix Pasila, Petra Christian University, Indonesia

IOP Conf. Series: Materials Science and Engineering **235** (2017) 011003 doi:10.1088/1757-899X/235/1/011003

Peer review statement

All papers published in this volume of *IOP Conference Series: Materials Science and Engineering* have been peer reviewed through processes administered by the proceedings Editors. Reviews were conducted by expert referees to the professional and scientific standards expected of a proceedings journal published by IOP Publishing.

IOP Conf. Series: Materials Science and Engineering **235** (2017) 012001 doi:10.1088/1757-899X/235/1/012001

Trade-off Analysis of Underwater Acoustic Sensor Networks

G Tuna[1,*] and R Das[2]

[1] Trakya University, Edirne, 22020, Turkey
[2] Fırat University, Elazığ, 23119, Turkey

* gurkantuna@trakya.edu.tr

Abstract. In the last couple of decades, Underwater Acoustic Sensor Networks (UASNs) were started to be used for various commercial and non-commercial purposes. However, in underwater environments, there are some specific inherent constraints, such as high bit error rate, variable and large propagation delay, limited bandwidth capacity, and short-range communications, which severely degrade the performance of UASNs and limit the lifetime of underwater sensor nodes as well. Therefore, proving reliability of UASN applications poses a challenge. In this study, we try to balance energy consumption of underwater acoustic sensor networks and minimize end-to-end delay using an efficient node placement strategy. Our simulation results reveal that if the number of hops is reduced, energy consumption can be reduced. However, this increases end-to-end delay. Hence, application-specific requirements must be taken into consideration when determining a strategy for node deployment.

1. Introduction

Underwater Acoustic Sensor Networks (UASNs) are used for various purposes in a variety of underwater applications, such as underwater data collection, coastal surveillance, oil monitoring, pollution detection, and seismic and volcanic prediction [1]. UASNs typically consist of two types of entities: nodes and a surface gateway to communicate to the external world. In UASNs, to measure different parameters such as conductivity, oxygen, pH, salinity, temperature, turbidity, and depth, sensors and probes are installed in the nodes and the nodes are deployed at the monitoring site. While, in Underwater Wireless Sensor Networks, the transmission of information between the sensor nodes can be realized in the form of acoustic, electromagnetic, or optical waves, acoustic communication is the communication solution in UASN applications [2].

Because of the nature of aquatic environments, there exist a variety of constraints, most importantly high propagation delay, high bit error rate (BER), low/medium bandwidth, and short-range communications [3]. These constraints both degrade the performance of UASNs and limit the lifetime of underwater sensor nodes, and hence call for the deployment of many underwater sensor nodes for successful UASNs. Although UASNs can be deployed using different approaches, it is a significant challenge because of the abovementioned constraints. In addition, approximate cost of underwater sensor nodes is much more than terrestrial wireless sensor nodes and this limits the number nodes to be deployed and eliminates some deployment approaches.

For the performance and success of UANs, deployment strategies play a key role since they need to address the major goals such as ensuring full-connectivity coverage and optimizing network lifetime. In addition, the deployment strategies must take into consideration major network and/or node-related factors including packet reception rate, delay, throughput, transceiver power, battery density, and CPU power [4]. Since UASN nodes are powered by internal/external batteries, they should be properly placed to minimize path loss and energy consumption. Typically, UASN nodes maintain their locations using their location and depth adjustment capabilities. Different types of manned or unmanned vehicles including boats, submarines, and aerial vehicles can be used for the initial deployment.

In this study we focus on analyzing the relations between energy consumption and number of hops and between end-to-end delay and number of hops for UASNs. The rest of the paper is structured as follows. The theory and results of a set of simulation studies carried out to analyze the effects of node placement on energy consumption and end-to-end delay are presented in Section II. Section III concludes the paper.

2. Effects of Node Placement on Energy Consumption and End-To-End Delay

In this section we present the details of a case study based on an orthogonal dam model and a simulation study to find out the effects of node placement on network energy consumption and end-to-end delay for the UASN. For the UASN, a sufficient number of underwater sensor nodes must be deployed to cover the volume of the dam model. Since the underwater sensor nodes communicate with each other using acoustic links, the path loss is caused by the energy spreading and the wave absorption [5]. In UASNs, mainly, while the former is dependent on transmission range [6, 7], the latter is frequency-dependent.

In the case study, two-dimensional (2D) acoustic communications were considered. Therefore, we assume that UASN nodes are equipped with the same model transceiver and cylindrical-based communication capability with a constant radius of d. Under this assumption, two underwater sensor nodes can communicate if the distance between the nodes is $\leq d_{\max}$. Moreover, all the nodes transmit with the same transmission power. In addition, all of the environmental parameters apply to all the nodes.

Let the UASN node n_i be located at a distance of d from the gateway and send its data packets to the gateway over a multiple-hop path $h_1h_2....h_N$ [8]. For the node n_i, the source signal level SSL which depends on the transmission power TP of the transceivers can be expressed by equation (1) [5, 7, 9].

$$SSL = ANL + SNR + TL - DI \tag{1}$$

where ANL represents the ambient noise level, SNR represents the signal-to-noise ratio at the transceiver, TL represents the transmission loss, and DI represents the directivity index. ANL increases in shallow water and depends on frequency and many other factors such as waves, wind, and surface-level activities. In the case study, ANL was assumed to be 50 dB based on the measurements in the literature [6]. Under the assumption of omnidirectional hydrophones, DI is assumed to be zero. Therefore, SSL can be expressed as a function of TL and SNR (2) and TL can be approximated as using equation (3) [6, 8]. SSL can also be expressed as the transmitted signal intensity at depth of 1 m using equation (4).

$$SSL = TL + SNR + 50 \tag{2}$$

$$TL = 10\log d + ACd \times 10^{-3} + TA \tag{3}$$

where d represents the distance between the sender and receiver in meters, AC represents the absorption coefficient, and TA represents the transmission anomaly.

$$SSL = 10\log\frac{SIT}{1\mu Pa} \tag{4}$$

where SIT represents the intensity of underwater signal in μPa and can be expressed using equation (5). To achieve the intensity SIT at a distance of 1 m from the sender to the receiver, the transmission power TP needed can be expressed using equation (6) [6, 8].

$$SIT = 10^{SSL/10} \times 0.67 \times 10^{-18} \tag{5}$$

$$TP = 2\pi \times 1m \times h \times SIT \tag{6}$$

where h represents the water depth in m.

In the case study, Orthogonal Frequency Division Multiplexing (OFDM) encoding and 16-Quadrature Amplitude Modulation (QAM) was assumed [10, 11]. In the UASN, the end-to-end delay EED can be expressed using equation (7) [8, 11].

$$EED = N\left(t_{prop(i,i+1)} + t_{pkt}\right) \tag{7}$$

where $t_{prop(i,i+1)}$ is the propagation time between underwater sensor nodes i and $i+1$, t_{pkt} is the time required to transmit a data packet, and N is the number of hops. The time required to transmit a data packet can be expressed using equation (8) and the propagation time can be expressed using equation (9).

$$t_{pkt} = PL / DR \tag{8}$$

$$t_{prop} = d / PV \tag{9}$$

where PL represents packet length, DR represents data rate, and PV represents the propagation velocity of the acoustic waves in water.

In the case study, the effects of placing two or more intermediate hops on the energy consumption and the end-to-end delay of the UASN were investigated. Based on the parameters [12, 13] listed in Table 1, a set of simulation studies was realized. Figure 1, Figure 2, and Figure 3 show energy consumption while sending one data packet. In the case study, distances between nodes of 125 m, 250 m, and 500 m were considered. The results indicate that when the number of hops is increased, the energy consumption decreases. Figure 4 shows the end-to-end delay as a function of the number of hops for distances between nodes of 125 m, 250 m, and 500 m. It can be seen that the end-to-end delay increases when the number of hops is increased. The results reveal that the number of hops should be correctly determined because there is a significant trade-off between UASN energy consumption and end-to-end delay.

Table 1. Simulation parameters

Parameter	Value
Temperature	15°C
Frequency	40 KHz
Absorption Coefficient	8.96 dB/km
Maximum Depth (Height)	100 m
Transmission Anomaly	5 dB
Propagation Velocity	1466 m/s at 15°C [13]
Packet Length	512 bits
Data Rate	1 Kbps
Noise Bandwidth	1 kHz
BER	10^{-9}

IOP Conf. Series: Materials Science and Engineering **235** (2017) 012001 doi:10.1088/1757-899X/235/1/012001

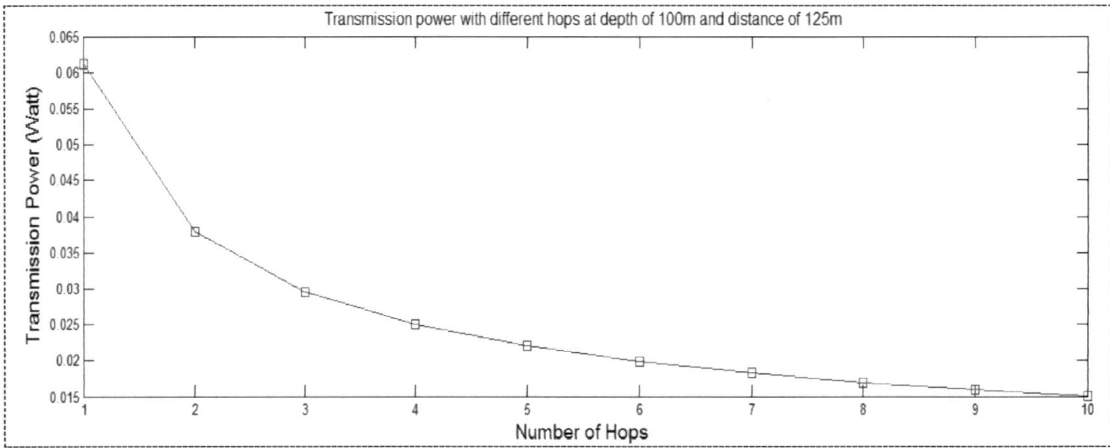

Figure 1. Energy consumption while sending one data packet at depth of 100 m. Transmission power with different hops at depth of 100 m and distance of 125 m.

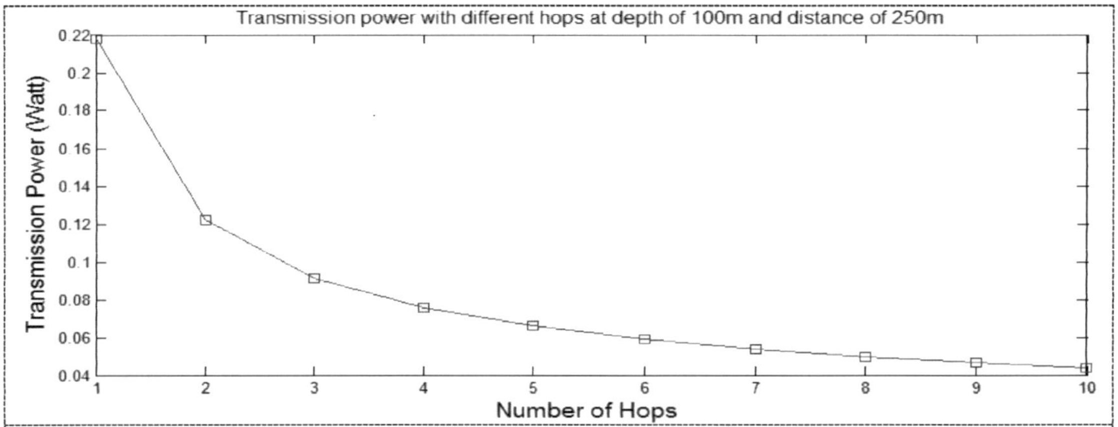

Figure 2. Energy consumption while sending one data packet at depth of 100 m. Transmission power with different hops at depth of 100 m and distance of 250 m.

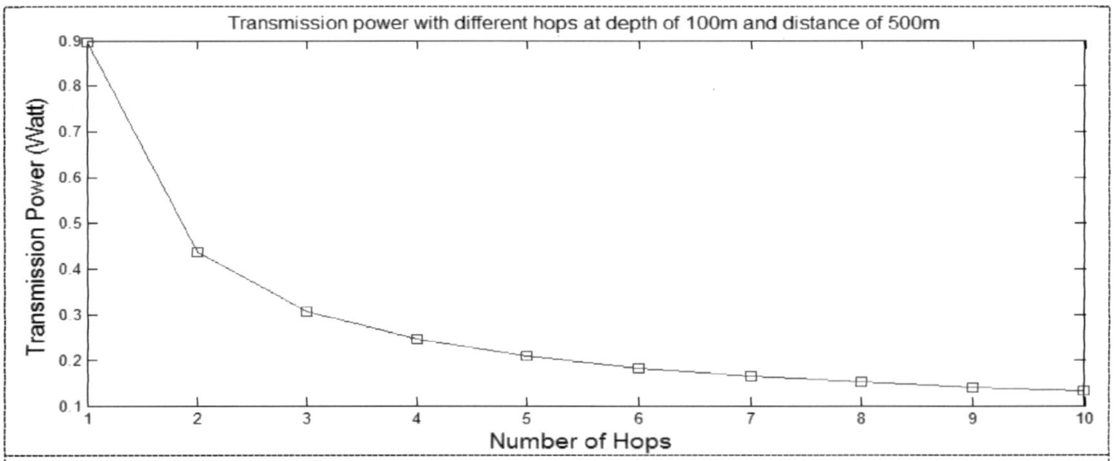

Figure 3. Energy consumption while sending one data packet at depth of 100 m. Transmission power with different hops at depth of 100 m and distance of 500 m.

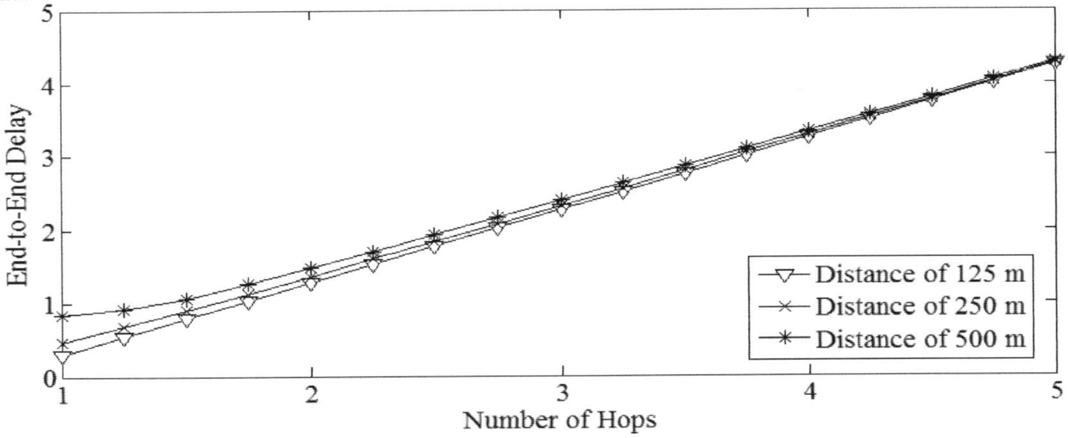

Figure 4. End-to-end delay of the UASN as a function of the number of the hops for different distances between nodes.

3. Conclusion

Although there is an increasing interest in UASNs due to their unique capabilities, the efficiency of UASNs and the lifetime of sensor nodes are greatly affected by a number of constraints including short-range communications, long propagation latency, limited bandwidth capacity, and variable bit error rate. To address these issues efficiently, the use of field experiments and test beds is essential and supports more accurate performance analysis and system characterization. Such work can increase overall robustness in different conditions and enable the analysis of total system cost and energy requirements. On the other hand, investigating and handling these issues are difficult due to the high cost and complexity of trials and field experiments, emulation and hardware-in-the-loop technologies, and physics-based simulation which should be used to estimate performance and must be designed to involve the presence of shadow zones, jamming, and natural interference. Due to the unavailability of such environments, in this paper, with a set of performance evaluations, the trade-offs between end-to-end delay, UASN energy consumption and number of hops was investigated to find out the optimal number of hops which satisfies both end-to-end delay and overall network energy consumption requirements. Our simulation results revealed that increasing the number of hops between the sender and receiver reduces energy consumption significantly but increases end-to-end delay.

References

[1] Heidemann J, Stojanovic M and Zorzi M 2012 Underwater Sensor Networks: Applications, Advances, and Challenges *Philosophical Transactions of the Royal Society-A* **370**(1958) 158-75
[2] Vasilescu I, Kotay K, Rus D, Dunbabin M and Corke P 2005 Data collection, storage, and retrieval with an underwater sensor network *Proc. the 3rd international conference on Embedded networked sensor systems* (New York) pp. 154-65.
[3] Lanbo L, Shengli Z and Jun-Hong C 2008 Prospects and problems of wireless communication for underwater sensor networks *Wirel. Commun. Mob. Comput*. **8**(8) 977-94
[4] Akyildiz I F, Pompili D and Melodia T 2005 Underwater acoustic sensor networks: research challenges *Ad Hoc Networks* **3**(3) 257-79
[5] Michailidis E T, Tuna G, Gezer G, Potirakis S M and Gulez K 2014 ANN-Based Control of a Multi-Boat Group for the Deployment of an Underwater Sensor Network *International Journal of Distributed Sensor Networks* **2014** 786154
[6] Urick R J 1983 *Principles of Underwater Sound* (Mcgraw-Hill)

[7] Stojanovic M and Preisig J 2009 Underwater acoustic communication channels: Propagation models and statistical characterization *IEEE Communications Magazine* **47**(1) 84-9

[8] Felemban M and Felemban E 2013 Energy-delay tradeoffs for Underwater Acoustic Sensor Networks *Proc. First International Black Sea Conference on Communications and Networking* (Batumi, Georgia) pp. 45-9

[9] Proakis J G 1995 *Digital communications* (3rd ed.) (New York: McGraw Hill)

[10] Farrell A C and Peng J 2012 Performance of IEEE 802.11 MAC in underwater wireless channels *Procedia Computer Science* **10** 62-9

[11] Cui S, Madan R, Goldsmith A and Lall S 2005 Energy-delay tradeoffs for data collection in tdma-based sensor networks *Proc. IEEE International Conference on Communications* (Seoul, Korea) **5** pp. 3278-84

[12] Fisher F H and Simmons V P 1977 Sound Absorption in Sea Water *Journal of Acoustical Society of America* **62**(558)

[13] Marczak W 1997 Water as a standard in the measurements of speed of sound in liquids *Journal of the Acoustical Society of America* **102** 2776

IOP Conf. Series: Materials Science and Engineering **235** (2017) 012002 doi:10.1088/1757-899X/235/1/012002

Strength Analysis and Reliability Evaluation for Speed Reducers

Yuo-Tern Tsai[1*] and Yung-Yuan Hsu[2]

[1]Department of Mechanical Engineering, DeLin Institute of Technology, 1, Lane 380, Ching Yun Road, Tucheng, Taiwan, 236, R.O.C.
[2]Department of Mechanical Engineering, Chung Hua University, Hsin Chu, Taiwan, R.O.C

*Corresponding author: yttsai@dlit.edu.tw

Abstract. This paper studies the structural stresses of differential drive (DD) and harmonic drive (HD) for design improvement of reducers. The designed principles of the two reducers are reported for function comparison. The critical components of the reducers are constructed for performing motion simulation and stress analysis. DD is designed based on differential displacement of the decelerated gear ring as well as HD on a flexible spline. Finite element method (FEM) is used to analyze the structural stresses including the dynamic properties of the reducers. The stresses including kinematic properties of the two reducers are compared to observe the properties of the designs. The analyzed results are applied to identify the allowable loads of the reducers in use. The reliabilities of the reducers in different loads are further calculated according to the variation of stress. The studied results are useful on engineering analysis and reliability evaluation for designing a speed reducer with high ratios.

1. Introduction

Reducers are extensively applied for speed reduction in many types of machinery such as machine tools, transmission mechanisms, and robots, etc. Their functional properties always are emphasized for satisfying the performance needs, such as speed-reduction ratio, transmitted torque, and efficiency, etc. An often used reducer for high reduction ratio and small volume is harmonic drive (HD). The HD is designed using a Flexible spline which possesses the properties of small backlash, high transmission accuracy, and large reduction ratio so that it is extensively used in robotic devices [1]. The existed problems of the HD are their rigidity due to the flexible design of the decelerated gear rings [2]. The other one frequently applied is trochoidal gear reducer which has the advantages of high torsional rigidity, great shock-resistant ability and loading capacity. [3]. A new design of reducers for high rigidity, large reduction ratios is to make use differential displacements of decelerated gear rings, named as differential drive (DD) [4]. This DD is designed primarily consisting of one off-center cam and several gear rings. Different types of reducers have been proposed in the past. However, many structural problems such as motion property, induced stress, and reliability, etc., are still unknown so that the further studies are needed to avoid failures.

1.1 Stress analysis

The failures of gear-based reducers frequently occur at the decelerated gears, for example, gear tooth breakage, pitting or wearing of tooth surfaces. Gear tooth breakage is caused by internal stress rupturing which is a direct result of the accumulated residual stress or root bending stress exceeding the strength of materials [5]. To preliminarily investigating the strength of a reducer, the working stresses must be observed. An often used approach in analyzing structural stresses and dynamic properties is by finite element method (**FEM**). The **ANSYS** software is a powerful tool of **FEM**, which has been extensively applied in many engineering fields. The geometric models created by **CAD** are imported into **ANSYS** for meshing and computing while performing statistic or dynamic analyses. It is sometimes difficult to obtain an optimal solution by **FEM** because the definitions may slightly violate the constraints, although the efficiency can be improved by replacing the search algorithms. Recently, the use of **FEM** to perform engineering stress analysis and support product design has been extensively studied. For example, Tsai et al. [6] made use of the limited experimental data and simulated stress obtained by **FEM** to evaluate the fatigue life of dental implants. A typical example reported by Tsai [7] is to integrate probabilistic theories with **CAD/CAE** to perform reliability optimization design for robotic arms.

1.2. Brief contents

This paper reports the design concepts of a high-ratio speed-reducer. Geometric models of the reducer are designed using parametric methods so that the related components can be rapidly created by changing the designed variables. The decelerated gear rings play a critical role in determining the ratios of speed reduction and the strength of the reducers. The structural stresses of the decelerated gear ring within the DD and HD are studied to formulate the relationship between the working loads and the tooth sizes so that the allowable loads can be decided accordingly. The strength and reliability of the two reducers are compared to assess the advantages of the designs. The **FEM** is used to evaluate the maximum stresses and deformations of the reducer under various loadings. The studied results are useful for stress analysis and reliability design in developing a reducers owning high reduction ratios.

2. Differential Drive (DD)

The working principles of the reducer design are based on a differential displacement of the sliding gear ring related to the fixed gear ring. The movements are similar to the sliding meter to the fixed meter in the traditional calipers. The critical components of the DD have four, the fixed ring, sliding ring, medium ring and off-center cam. The isotropic and plain perspectives of the DD are illustrated in Fig.1.

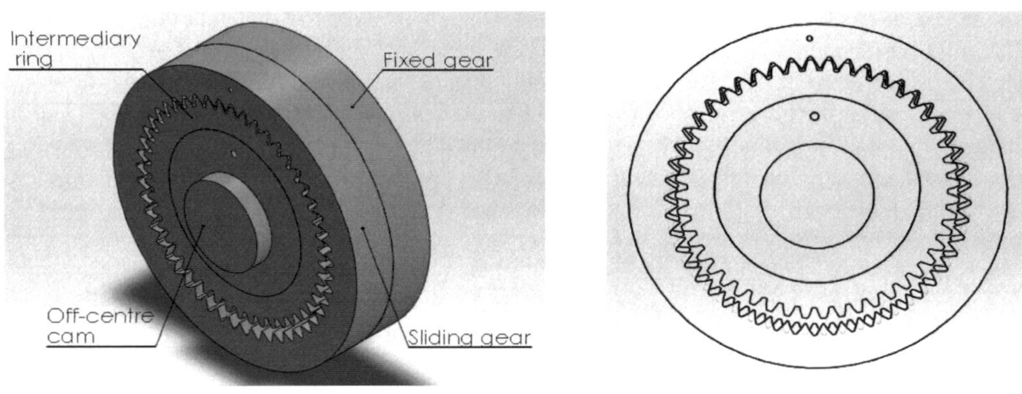

(a) isotropic view (b) plain view

Figure 1. Assembly drawing of the DD

The fixed ring is designed to guide the moving of the medium ring as a planetary gear. The medium ring is designed to receive the power inputted from the off-center cam. The medium ring would generate planetary motion and push the sliding ring to move while the off-center cam is rotating. The fixed and sliding rings are designed with different teeth so that a slight difference of position of the two paired teeth can be generated. The paired teeth are forced by the teeth of the medium ring to align with each other. The differential displacement generated by the fixed and sliding rings would result in a large speed reduction of the sliding ring. The ratios of speed-reduction are dependent upon the number of tooth of the fixed and sliding rings.

The ratio of speed reduction can be evaluated by

$$s = \frac{z_s}{z_s - z_c}, \tag{1}$$

where z_s, z_c stand for the number of teeth on the sliding and fixed rings, respectively. The off-center cam is designed according to the sizes of the paired gear rings. The off-center distance is expressed as

$$e = \frac{D_c - D_s}{2}, \tag{2}$$

where D_s, D_c are the pitch diameters of the sliding and the fixed gear rings, respectively.

Possible failure modes of the DD will occur at the medium ring because it is the weakest one of the structure. They are likely to appear depending on the pitch-line velocity and transmitted torque. The failures occur at the positions of engagement of the gear teeth include both tooth breakage and tooth surface pitting. Breakage of the tooth is caused by the root bending stress exceeding the strength of the materials. Pitting of tooth surface is incurred by repeated surface contact stress as well as inadequate surface strength. Increasing the transverse contact ratio as well as enlarging the overlap between the tooth contact-pairs are effective approaches to reducing noise and vibration.

The horizontal force can be obtained by taking the force equivalence [4] as follows:

$$F_2 = \frac{T_1}{(r_2 + e)}. \tag{3}$$

The moment acting on the intermediary ring can be obtained by equivalent equivalence on acting and reacting forces as

$$T_2 = \frac{r_2}{(r_2 + e)} T_1. \tag{4}$$

The structural stress on the gear rings can be further evaluated according to the results of loading analysis, based on the theory of gear design. Aiming to the paired transmission of the inner gears, the induced stress including the Root Bending Stress (**RBS**) and Surface Contact Stress (**SCS**) can be evaluated according to the transmitted torque T_2. Generally, the **RBS** is commonly evaluated based on the Lewis equation. The gear tooth is modeled as taking the full load at its tip as with a simple cantilever beam. The **SCS** is commonly predicted based on Hertz theory in which the contact points are simulated as two contacting cylinders. In this paper, the structural stresses are obtained by **FEM**.

3. Harmonic drive (HD)

Harmonic drive is extensively applied in industrial robots. It is designed with toothed mechanism which is mainly composed by three elements as shown in Fig. 2 [8]:

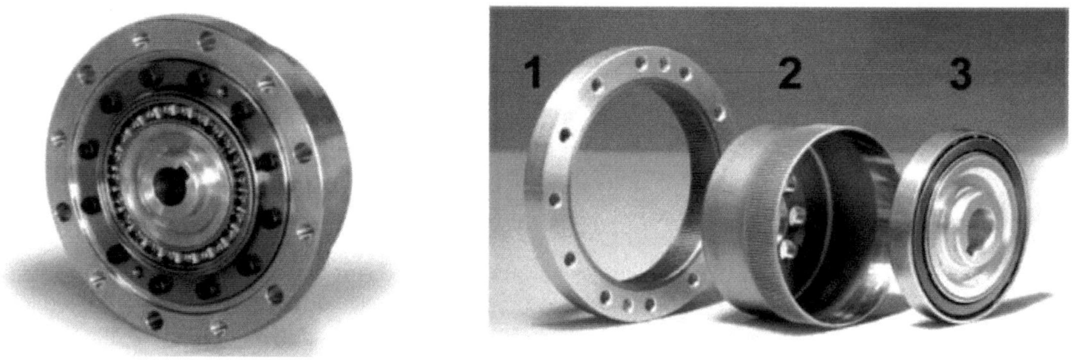

Figure 2. Basic components of the harmonic drive

The designed properties of the HD are introduced in following:

- The FS is designed slightly smaller in diameter than the CS and the teeth of the FS usually are two fewer than the CS. The WG is designed with elliptical shape so that the FS is fitted as an ellipse and the teeth can engage the CS at two regions at the opposite ends of the major axis of the ellipse FS.
- The WG enforces the FS for gear tooth engagement at the major axis of the ellipse and is acting as an input rotator of power. The WG clockwise movement for each 180°, the FS would move counterclockwise by one tooth relative to the CS (fixed).
- The FS would move counterclockwise by two teeth from its previous position relative to the CS for each complete clockwise rotation of the WG.

The FS is deflected as elastically deformation while the WG is elliptically across the major axis. The teeth of the FS are engaging simultaneously with the ring gears of the CS at two zones of either end of the major elliptical axis. On the other hand, there is no engagement arisen when the elliptically deflected FS across the minor axis. The meshing zones of the FS rotate with the generator when the WG rotates. A difference of two teeth of the tooth number of the FS and the CS results in a relative movement of the two wheels when the FS is driven by the WG. The FS moves relative to the CS having an angle equivalent to two teeth after a complete rotation of the WG.

HD is currently used in many automotive and space industries including aviation, medicine, automatics and robotics. HD has numerous advantages in transmission compared to the classical gear chain. The main advantages are high single-stage reduction ratios including compact design, high torque capacity and high torsion stiffness, zero backlash, excellent positioning accuracy and repeatability, etc. Moreover, the drawbacks are high elasticity and nonlinear stiffness and damping. The application of harmonic drives in various industrial fields is more and wider due to the excellent performances. The disadvantages of HD is the strength problems due to the deflect design of the FS. The FS is the main component of a harmonic drive, which can generate a repeated vibration by the wave generator. With this a reason, the FS should have flexibility and good vibration characteristics.

A simplified model of the HD is constructed for analyzing the structural stresses. The geometric size and the speed-reduction ratios are designed as well as the ones of the DD for convenient in strength comparison. The tooth's shapes are constructed based on a combination of different curve profiles which is similar to a commercial product. The geometric models including the CS, FS and WG of HD are created for structural analysis. This means that inner diameter of the FS is about 48 mm and the tooth's numbers for the FS and the CS are 100 and 102, respectively.

4. Strength analysis

ANSYS is a tool of **FEM** for obtaining the simulated features of a design in real world by using mid-plane extraction of structures, boundary conditions, mesh controls, loads and perimeter welds, etc. The geometric models produced by **CAD** software are imported into **ANSYS** to analyze the static and dynamic properties. For gear transmission, the bending and contact stresses between the mated teeth

IOP Conf. Series: Materials Science and Engineering **235** (2017) 012002 doi:10.1088/1757-899X/235/1/012002

are the studied topics. Here, the induced stresses of the medium rings of the DD and the FS in the HD are the concerned problems since they are the weakest ones within the reducers. The analyzed results are further used to evaluate the strength and reliability of the reducers.

The material of the DD is designed with structural steel. To simplify the analyzed model, only the critical components, the inner-outer gear rings of the DD, are loaded into **ANSYS** to perform stress analysis. The augmented Lagrangian formulation method is selected as the most suitable for the analysis from among the available nonlinear methods because it is best for frictionless contact. The connection of the inner-outer gear rings are set to frictionless contact and the cam is bonded to the inner ring. A fixed support is placed at the outer gear ring as well as a frictionless support at the input hole of the off-center cam. The gear thickness is set to 10 mm and the loading is set to be 100 Nm.

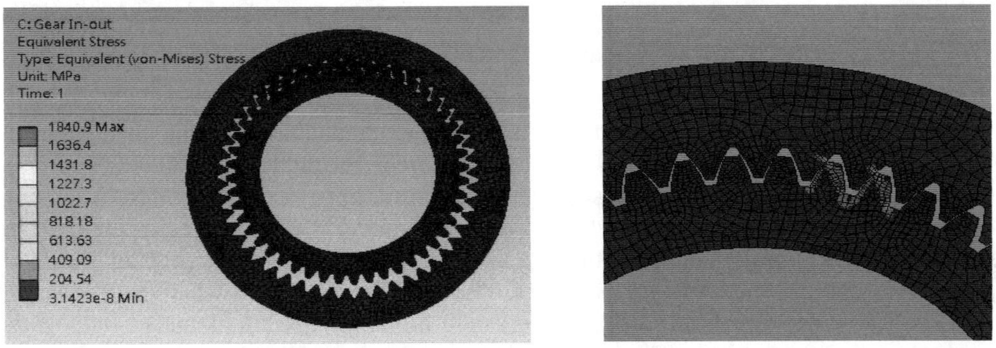

Figure 3. The surface contact stress

Here, the interval of the mesh is set to 100 and the meshes in the areas of the mating are strengthened by setting a smaller meshing size. The analytical results for equivalent stress as well as the contact stress are shown in Fig. 3. The maximum value of the **SCS** for this example is **1840 MPa,** occurring at the contacting position of the engaging teeth. The principle stresses, as well as the bending stress, are shown in Fig. 4. The maximum value of the **RBS** is **654 MPa,** occurring at the tooth roots. The evaluated results for maximum **RBS** and **SCS** are similar to the values obtained by numerical calculations. The results imply that applying the **FEM** to evaluate the **RBS** and **SCS** of the reducer is feasible.

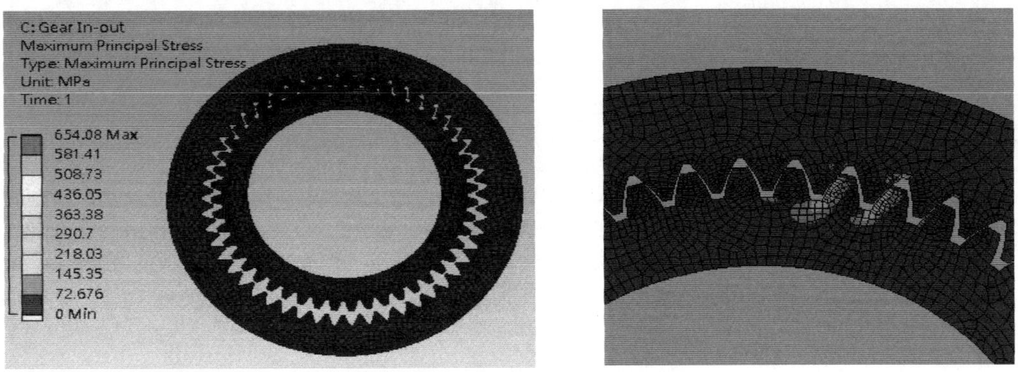

Figure 4. The root bending stress

The analyzed results show that the **SCS** is about 3 times of the **RBS**. Normally, the results of **FEM** analysis are closely related to the mesh settings. The convergence conditions of **FEM** analysis show that the maximum equivalent and principle stress will converge to one stable value representing the evaluations are reasonable.

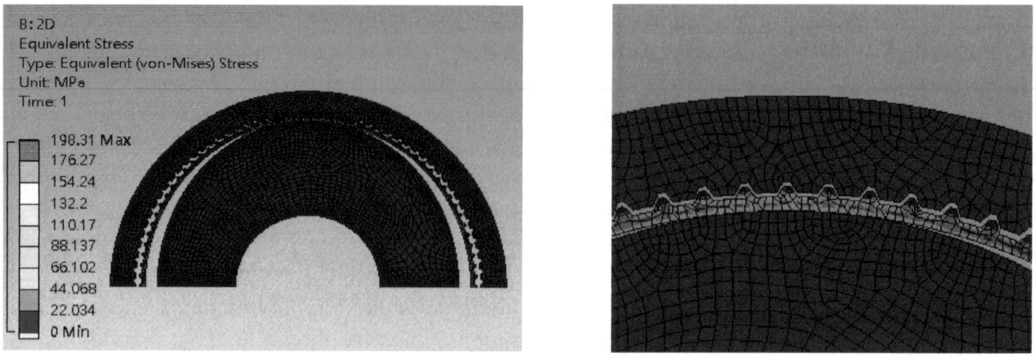

Figure 5. Equivalent stresses

On the other hand, the geometric models of the HD are also imported into ANSYS to perform stress analysis. The symmetrical geometries are used for simplifying the analyzed time. This is a nonlinear analysis since the FS and the WG existing an overlapping area. The CS is set to fixed support as well as the WG to frictionless support. The maximum stresses of the FS deformation caused by the WG are about 198 MPa as shown in Fig. 5. Moreover, the working stresses of the FS are obtained by setting the acting moments at the FS. The maximum principle stresses of the FS are about 1400 MPa while the loading moment is 100 Nm. The total stresses of FS would be about 1600 MPa. The analyzed results show that the working stresses of the DD are smaller than the ones of the HD at the same torques. It means that the allowable loads of the DD would be larger than the ones of the HD.

5. Reliability evaluation

Reliability is a probability of success of failure which is formulated based on **SSI** theory. Reliability index can be rated based on the second-order moment method while the strength random variable X is normally distributed with a mean value of μ_X and standard deviation of σ_X. On other hand, the stress random variable Y is also normally distributed with mean value and standard deviation (μ_Y and σ_Y.), respectively. The geometric distance indicating reliability index can be rated by

$$z = \frac{\mu_X - \mu_Y}{\sqrt{\sigma_X^2 + \sigma_Y^2}} \tag{5}$$

The relationship between reliability index, z, and the reliability can be evaluated by

$$R = 1 - \Phi(-z) = \Phi(z) \tag{6}$$

For example, reliability index z=3 represent the reliability being R= 0.99865.

Focusing on mechanical design, the reliability of a design can be expressed from the strength and stress distributions. If the strength information of materials (μ_X, σ_X) is known, the induced stress (μ_Y, σ_Y) would be the designed variants. Considering the coefficients of variation, the standard deviation of the induced stress can be evaluated by Eq.(7) when the designed variants γ_Y are known

$$\sigma_Y = \gamma_Y \mu_Y . \tag{7}$$

As a result, we can derive the equation of reliability design for the reliability target being set to R≥Φ(z).

Based on the evaluated approach, the reliabilities of the reducers at various loads can be further evaluated by integrating the analyzed results in **ANSYS** and the **SSI** theory. The induced stresses can be regarded as normal distributions while considering the variation of the design variables. In this paper, the strength variation of the materials is set to 10 percent of the mean values. The reliability of the reducers can be calculated by above equations in cooperation with the analyzed stresses in ANSYS. The reliability degradation with respect to the loading torques can be obtained. The analyzed results show that the reliability degradation is faster when the stress variation doesn't be considered. The analyzed results are useful for determining the allowed loads of the reducers.

6. Conclusions

The analytical methods of structural stress are reported for performing stress and reliability evaluations. **FEM** is used as a tool to analyze the stresses of the reducers at different designs and loads. The maximum stresses of the DD and the HD are identified to observing the most possible failed components. The analyzed results show that the induced stresses in DD are smaller than the ones in HD at the same loads. It means that the DD have better strength compared with the HD under the same designed conditions. According to the analyzed results, the allowable loads as well as the maximum moments of the reducers can be decided accordingly. The reliabilities of the reducers under different loads are evaluated by integrating the analyzed stresses with stress-strength interference theory. The studied results may be useful in designing, structural stress analysis and reliability evaluation for developing a high-performance reducer.

Acknowledgements

This work was supported by a grant from the National Science Council under contract No: MOST 105-2221-E-237-005. The authors would like to express their appreciation to the reviewers for their valuable suggestions which have helped to greatly improve the manuscript.

References

[1] Taghirad, H. D. and P. R. Belanger, 2009, "Modeling and parameter identification of harmonic drive systems", *Journal of Dynamic Systems, Measurement and Control*, Vol.120 (4), pp. 439-444.
[2] Ghorbel, F. H., P. S. Gandi and F. Alpeter, 2001, "On the kinematics error in harmonic drive gears", *Journal of Mechanical Design*, Vol. 123(1), pp.90-97.
[3] Shuting Li, 2014, "Design and strength analysis methods of the trochoidal gear reducers", *Mechanism and Machine Theory*, Vol. 81, pp.140-154.
[4] Y.T. Tsai and K. H. Lin, "Structural stress analysis and reliability evaluation for a new speed reducer", *Journal of Mechanics*, Published online: 13 September 2016, pp.1-10 (DOI: http://dx.doi.org/10.1017/jmech.2016.96).
[5] Alban L. E., 1993, "Systematic analysis of gear failures", *American Society for Metals*.
[6] Tsai, Y.T., K.S. Wang and J.C. Woo, 2013, "Fatigue-life and reliability evaluations of dental implants based on computer simulation and limited test data", *Proc. IMechE, Part C: J. Mechanical Engineering Science*, Vol.227 (3), pp. 554-564.
[7] Tsai, Y.T., K. H. Lin and Y.Y. Hsu, 2013, "Reliability optimization design for mechanical devices based on modeling processes", *Journal of Engineering Design*, Vol.24 (12), pp.849-863.
[8] General Catalogue Harmonic Drive AG, 05.2009.

IOP Conf. Series: Materials Science and Engineering **235** (2017) 012003 doi:10.1088/1757-899X/235/1/012003

Fault Tree Analysis for an Inspection Robot in a Nuclear Power Plant

Thomas A. Ferguson[1] and Lixuan Lu[1]

[1]Faculty of Energy Systems and Nuclear Science, University of Ontario Institute of Technology, 2000 Simcoe Street North, Oshawa, Ontario, Canada L1H 7K4

E-mail: thomas.ferguson1@uoit.net; Lixuan.Lu@uoit.ca

Abstract. The life extension of current nuclear reactors has led to an increasing demand on inspection and maintenance of critical reactor components that are too expensive to replace. To reduce the exposure dosage to workers, robotics have become an attractive alternative as a preventative safety tool in nuclear power plants. It is crucial to understand the reliability of these robots in order to increase the veracity and confidence of their results. This study presents the Fault Tree (FT) analysis to a coolant outlet piper snake-arm inspection robot in a nuclear power plant. Fault trees were constructed for a qualitative analysis to determine the reliability of the robot. Insight on the applicability of fault tree methods for inspection robotics in the nuclear industry is gained through this investigation.

1. Introduction

Most existing Nuclear Power Plants (NPPs) around the world are facing the challenges of aging equipment. As of 2013, 190 out of the 437 civil nuclear reactors have been in operation for more than 30 years [1]. However, the need of these power plants has increased over the past decade due to concerns over greenhouse gas emissions and the increasing demand for electricity. Therefore, the extension of the life expectancy of the aged/aging reactors has becoming more and more imperative [2].

In order to maintain safe and profitable operation of these life-extended reactors, important components have to be replaced. In cases where replacement is too costly or not feasible, the continual inspection and maintenance becomes paramount [3]. However, to limit the dosage exposure for workers, inspection can become difficult and costly. An attractive alternative for human inspection of critical components is the use of remotely operated robotics [3, 4].

Since the inception of the robotic industry, the major benefits of using robots in hazardous environments have become increasingly evident. This is particularly true in the nuclear industry, where robots have been used routinely [4, 5]. The versatility of the robotic technology has been fully exploited over the years. Common roles performed by robots include handling high level radioactive loads [6], decommissioning [4], testing and maintenance [3, 4], and assisting after nuclear disasters [6]. Although nuclear robotics are a well used technology, there is still continual development as more challengers are presented as the industry ages. One of the most prominent challenges is the increasing demand of inspection. Inspection robotics play an important role in monitoring and detecting defects in critical components so as to determine early warning of component failure and prevent possible nuclear accidents. It is essential to understand the reliability of the robots performing these required inspection tasks, so as to gain confidence in the inspection results. Furthermore, the need for clear understanding of the paths to failure is compounded by the fact that these inspection robots work in an ionizing radiation field and have limited amount of exposure before failure. A significant cause of malfunction of tele-operated robots in the nuclear industry is a result from radiation damage [5, 7]. This paper presents a reliability assessment of such robotics using the Fault Tree (FT) analysis technique, with emphasis on a case study of a snake-arm robot for inspection of primary heat transport pipes of a nuclear reactor.

2. Snake-Arm Pipe Inspection Robot

The pressurized coolant outflow pipes from the nuclear reactor are components that are not feasible for replacement, and with age must be monitored. These pipes become prone to thinning due to flow-assisted corrosion, after years of operation. Furthermore, due to the long length of the pipes structural support is required. The constant motion between the feeder pipes and the support structure caused outside pipe fretting, which further caused the thinning of the pipes. This could cause a compromised pressure boundary, which could result in the release of high pressure radioactive steam.

The potential of fretting requires frequent and continual external inspection to be conducted. The examination has historically been performed manually by visual inspection, where the operator uses a camera to capture data for further inspection. This procedure can only be performed during outages, and therefore, due to geometric and time restraints, some areas cannot be viewed and points of concern are difficult to revisit. To alleviate these problems a snake-arm robot with high resolution cameras attached was developed.

The inspection robot is a remotely operated machine with a 2.2 m long, 12.5 mm width, and 18 DoF snake arm attached to a remotely controlled vehicle base. At the tip of the arm is the main inspection tool equipped with two forward pointing and one sideways facing fixed focal length cameras. The cameras have high powered LED lights matched to each respective camera's focal point. The video captured by the cameras are sent to a workstation outside of the hazardous area, by an umbilical cable. The arm is divided into multiple sections, with multiple control cables attached at the end of each section. The arm is manipulated by actuators at the base of the arm, so no wires or electronics are in the arm itself. Tip-guiding software is used to operate the arm.

The implementation of the robot provides multiple benefits, which include reduced operator exposure, constant availability since inspection is not limited to outages, and examination of difficult-to-reach places. The robot itself is not a high risk device to damage any equipment because of its size and structure. However, if the robot was to fail in locating any defects, a significant nuclear incidence could occur [3], due to the consequence of inspection robotic not being able to complete its task. Therefore, the reliability of the robot must be assessed. Furthermore, the dominant cause for failure of components is from radiation damage. Although the radiation tolerance for components is well known [7, 8], the propagation of these failures should also be assessed for a strong understanding of the robotic system reliability.

3. Fault Tree analysis

Fault Tree Analysis (FTA) is a top-down deductive failure analysis, which illustrates the sequence of failures that could result in a system failure (top event). The methodology uses a graphical illustration of the failure events. A failure path is determine by combining failure events using the standard logic gates (AND, OR, etc.) in a tree like fashion [9]. Fault Tree analysis is a well defined and commonly used tool in many different industries. Common fields of use include aerospace, nuclear power generation, and other events that have a significant safety risk to human health. FTA has also become a common method for assessing the reliability and safety of robotic systems [10-15], including those used within the nuclear industry [5, 16]. For a more general review of reliability assessments for robotics please refer to references 13 and 14.

Since the snake-arm robot itself is not a high risk device, the top events for the construction of fault trees for analysis are determined by the events that would not allow the robot to inspect the feeder pipes and identifying defects leading to a compromised pressure boundary. The top events were evaluated to be arm deployment apparatus failure, arm control systems failure, and the inability to control the locomotion of the device. The top event with lowest risk tolerance was determined to be the arm control systems failure, and will be the focus of this study.

A number of assumptions were made to produce the fault trees. The design and function of the device are described in [3, 17, 18]. The references state that the device arm is controlled by multiple cables connecting to each segment of the arm. The actuators are described as electronic motors. Furthermore, each segment of the arm can be adjusted independently from other segments, so the failure of actuators of one segment does not lead to loss of control of the whole arm. As a result, it is assumed that the inability for inspection occurs with failure of all section actuators. Any adjustments to the above assumption could lead to variation in the fault trees.

4. Results and Discussion

As mentioned above the top event with the lowest fault tolerance was found to be the arm control system failure. The main reason for this is because for failure of the arm deployment apparatus to occur all sections of the arm are required to fail. Each segment of the arm is controlled in parallel, so if one arm segment were to fail the other arm segments can still mechanically be controlled.

Although the actuators for each segment of the arm are independent of each other, the control calculations do not share the same independence. The device uses a tip-guiding software to aid in the control of the arm by the operator. In order to determine the position of a segment the controller requires measurements both from the arm segment under control and measurements from the other segments. As a result, if there was a failure in the measurement in one section of the arm all other sections will be affected. Therefore, if a fault was to occur lower down on the tree it will propagate up to the top event, as demonstrated by the OR-gates present in the tree. The constructed fault tree is shown Figure 1 below.

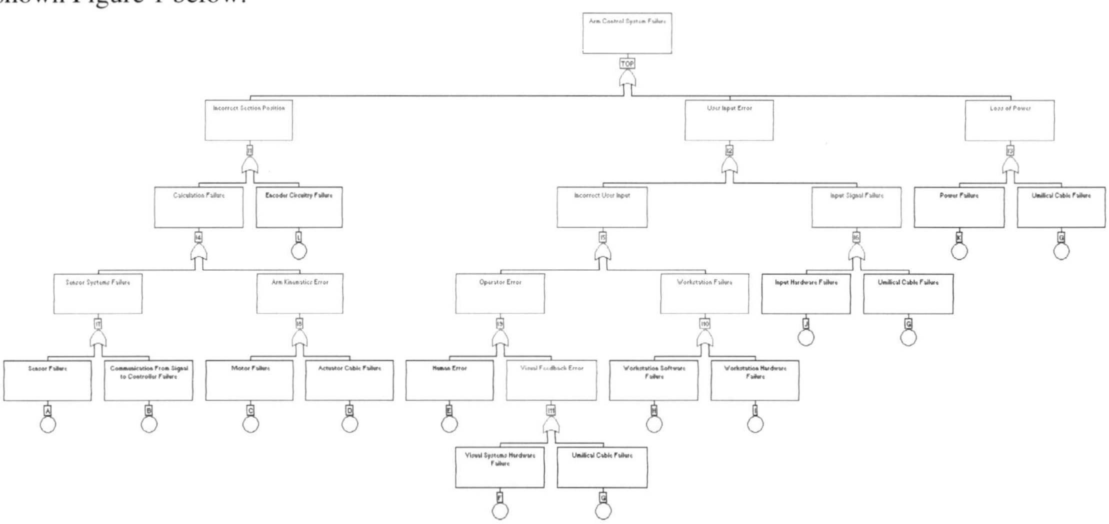

Figure 1. Fault tree of arm control system failure.

As can be seen from Fig. 1, the fault paths resulting in arm control system failure can be clearly extracted from the fault tree. From the failure paths, the effect of radiation susceptible component malfunctions can be determined. Components and materials that are susceptible are well known in literature, examples include: electronics, sensors, fibre optics, insulating materials and lubricants [7, 8]. From this knowledge the initiating events that are caused by dosage limited component failure were determined.

After the construction of the tree, the number of minimum cut-sets was determined. There were a total of 12 minimum cut-sets of the first order, and of the 12 minimum cut sets, 7 of them are susceptible to radiation damage. The initiating events with the lowest limit of exposure are:

communication from sensors to controller failure, visual systems hardware failure, input hardware failure, and encoder circuitry failure. Since the system has a low fault tolerance, and any failure will propagate up the tree, dosage limits should be known and monitored. Radiation hardened components and materials should be used. However, they do not grant indefinite exposure and failure paths are still important to know.

Improvements can be made to reduce the fault propagation in the system. Suggested improvements to be made include the following: An increase in the number of sensors would reduce the risk of sensory failure. To take full advantage of the redundant sensors, a sensor fault detection and reconfiguration algorithm implemented in the arm controller would be necessary [19]. The purposes of the algorithm would be to detect a faulty sensor, exclude its respective signal in control calculation and include only the remaining functional sensors. In order to take advantage of the multi-redundant manipulator, the control software must be able to adjust the kinematics of the arm after a segment failure. A method to achieve such flexibility is by having multiple kinematic calculations for possible section failure present in the controller, and a reconfiguration algorithm to correct the controller to the appropriate calculation after failure [20].

The fault tree analysis presented in this study was an entirely qualitative study due to limitations of available data. However, the purpose of this study was to demonstrate fault tree analysis as a tool to understand the reliability of nuclear inspection robots and determine the areas of most concern in terms of radiation damage. The ability to identify these areas is very useful for both operation and design phase of the robotic systems life cycle. Operators are able to determine which components require the most detailed dosage monitoring, and formulate corresponding mitigation strategies. The understanding of low fault tolerance of radiation damage areas of concern determined from the FTA, can allow for identifying improvements required, and what components require radiation tolerant hardware while designing the robot.

5. Conclusion

The reliability of nuclear inspection robotics was investigated by fault tree analysis with emphasis on a case study of a pipe inspecting snake-arm robot in a nuclear reactor. The top event for the FTA with the lowest fault tolerance was found to be failure of the arm control system. A significant source of component malfunction of nuclear robotics is from radiation damage. The fault tree analysis allowed for determination of failure paths for malfunctions of radiation damage susceptible components, which allows identification of areas of concern for possible improvement and formulation of mitigation strategies. The insights gained through the fault tree analysis of the snake-arm pipe inspection robot in this paper are beneficial for both design and operating phases of the system's life cycle.

References

[1] Schneider, M., & Froggatt, A. (2013). World nuclear industry status report 2013
[2] León, P. T., Cuesta, L., Serra, E., & Yagüe, L. (2010). Life Extension of Nuclear Power Plants: World Situation and the USA Case. *Revue générale nucléaire*, **6**, 50-56.
[3] Buckingham, R., & Graham, A. (2012). Nuclear snake-arm robots. *Industrial Robot: An International Journal*, **39(1)**, 6-11.
[4] Bogue, R. (2011). Robots in the nuclear industry: a review of technologies and applications. *Industrial Robot: An International Journal*, **38(2)**, 113-118
[5] Lauridsen, K., Christensen, P., & Kongsø, H. E. (1996). Assessment of the reliability of robotic systems for use in radiation environments. *Reliability Engineering & System Safety*, **53(3)**, 265-276.
[6] Kawatsuma, S., Fukushima, M., & Okada, T. (2012). Emergency response by robots to Fukushima-Daiichi accident: summary and lessons learned. *Industrial Robot: An International Journal*, **39(5)**, 428-435.
[7] Sharp, R., & Decreton, M. (1996). Radiation tolerance of components and materials in nuclear robot applications. *Reliability Engineering & System Safety*, **53(3)**, 291-299.

[8] Houssay, L. P. (2000). *Robotics and radiation hardening in the nuclear industry* (Doctoral dissertation, University of Florida).

[9] Xing, L., & Amari, S. V. (2008). Fault tree analysis. In *Handbook of Performability Engineering* (pp. 595- 620). Springer London.

[10] Karbasian, M., Mehr, M. R., & Agharajabi, M. (2012). Improvement of the Reliability of Automatic Manufacture Systems by Using FTA Technique. *International Journal of Industrial Engineering*, **23(2)**, 85-89.

[11] Winfield, A. F., & Nembrini, J. (2006). Safety in numbers: fault-tolerance in robot swarms. *International Journal of Modelling, Identification and Control*, **1(1)**, 30-37.

[12] Yakymets, N., Dhouib, S., Jaber, H., & Lanusse, A. (2013, November). Model-driven safety assessment of robotic systems. In *Intelligent Robots and Systems (IROS), 2013 IEEE/RSJ International Conference* on (pp. 1137-1142). IEEE.

[13] Khodabandehloo, K. (1996). Analyses of robot systems using fault and event trees: case studies. *Reliability Engineering & System Safety*, **53(3)**, 247-264.

[14] Visinsky, M. L., Cavallaro, J. R., & Walker, I. D. (1994). Robotic fault detection and fault tolerance: A survey. *Reliability Engineering & System Safety*, **46(2)**, 139-158.

[15] Dhillon, B. S., Fashandi, A. R. M., & Liu, K. L. (2002). Robot systems reliability and safety: A review. *Journal of quality in maintenance engineering*, **8(3)**, 170-212.

[16] Walker, I. D., & Cavallaro, J. R. (1996). Failure mode analysis for a hazardous waste clean-up manipulator. Reliability Engineering & System Safety, 53(3), 277-290.

[17] Graham, A. C., & Buckingham, R. O. (2014). U.S. Patent No. 8,635,928. Washington, DC: U.S. Patent and Trademark Office.

[18] Buckingham, R. O., & Graham, A. C. (2012). U.S. Patent No. 8,126,591. Washington, DC: U.S. Patent and Trademark Office.

[19] Visinsky, M. L., Cavallaro, J. R., & Walker, I. D. (1994). Robotic fault detection and fault tolerance: A survey. *Reliability Engineering & System Safety*, **46(2)**, 139-158.

[20] Lewis, C. L., & Maciejewski, A. A. (1994). Dexterity optimization of kinematically redundant manipulators in the presence of joint failures. *Computers & electrical engineering*, **20(3)**, 273-288.

IOP Conf. Series: Materials Science and Engineering **235** (2017) 012004 doi:10.1088/1757-899X/235/1/012004

A Direct Proof of Significant Directed Random Walk

Choon Sen Seah[1, a], Shahreen Kasim[1, b], Mohd Farhan Md Fudzee[1, c], and Mohd Saberi Mohamad[2, d]

[1]Soft Computing and Data Mining Centre, Faculty of Computer Sciences and Information Technology, Universiti Tun Hussein Onn Malaysia
[2]Faculty of Computing, Universiti Teknologi Malaysia, Skudai

[a]hi150021@siswa.uthm.edu.my, [b]shahreen@uthm.edu.my, [c]farhan@uthm.edu.my, [d]saberi@utm.my

Abstract. This paper is presented to disclose the relationship between weight and connectivity of nodes. An equation is formed to enhance the connectivity of nodes in directed graph via weigh. With implementation of references data, the adjacency matrix is further enhances to increases the accessibility of nodes via vector. The evolution of random walk is disclosed in this paper as well. Significant directed random walk will be used to prove the importance of weight in this paper.

1. Introduction

The basics of random walk can be traced back from a famous study by Brown (1828), known as Brownian motion [1]. Random walk was first developed to predict the mosquito infestation in forest by Karl Pearson [2]. Nowadays, random walk is applied in many different fields for prediction purposes and hence, it was further developed to improve the accuracy of prediction. Random walk was developed into directed random walk where it is classifying under biased random walk [3]. Directed graph consists of a sequence of vertices with connection of edges [4]. With a specific vector, an initial vertex will be transported to another vertex via edges. Implementation of directed graph will be further discussed on section 3 and 4. The connection between vertices are improved with the implementation of weight during the calculation. The equations are further proved through implementing numeric values of gene expression data after data pre-processing via Gene Chip Robust Multiarray Averaging (GCRMA). The comparison of results between directed random walk and significant directed random walk are shown in section 5.

2. Random Walk

Random walk can be described as the movement of a particle in a certain state space under the random action [5]. The state space is usually a dimensional Euclidean space or the integral lattice. Furthermore, random walk was then drawn to the subject and many important fields, such as random processes, random noise, spectral analysis and stochastic equations [3]. Each step of random walk is either equal to +1 (step forward) or -1 (step backward).

3. Directed Random Walk

The Directed Random Walk (DRW) developed by Liu have the ability to restart and stimulate a random walker that starts on a source node, s [6]. At every time step, the walker transit from its current node to another randomly selected neighbour (forward) or goes backward to source node *s* with probability *r*. Formally, the DRW with restart is defined as

$$W_{t+1} = (1-r)M^T W_t + rW_0 \qquad (1)$$

where W_t is a vector which the i node holds the probability of being at node *i* at time, *t*. *M* is the row-normalized adjacency matrix of the graph, G. When the random walk begins, the initial probability vector, W_0 was constructed by assigning to each node whose initial probability was 0. W_0 is absolute t-test score, which will be further normalized into a unit vector [6]. The restart probability *r* was set as 0.7. W_t converges to a unique steady state in the presence of the ground node. This was obtained by performing the iteration until the normalization fall between W_t and $W_{t+1} < 10^{-10}$.

4. Significant Directed Random Walk

In this section, an improved DRW which named as significant directed random walk (sDRM) is presented. The sDRW proved that the weight of the nodes can affect the connectivity of nodes, which leads to higher vector. Directed graph is defined as weighted graph when there are values attached to the directed edges [7]. These values represent the cost of travelling from one node to the other. The cost can be measure in many terms, depending on the application.

For example, distance between two nodes and the average travelling time in minutes. In the case of gene classification, weight of gene is used as a parameter to identify the usefulness in classification [8]. The significant genes normally have higher weight compared to the others. This is because of the common genes across protein-protein interaction network. Weight of genes can be obtain from gene expression data.

The Significant Directed Random Walk have ability to enhance the connectivity between nodes by weight of nodes. At every time step, the walker transitions from its current node to a randomly selected neighbour (based on edge weights) or goes back to source node, s with probability *r*. *r* can vary according to the datasets due to the attraction of nodes [9]. For example, *r* can be 0.1, 0.2, 0.3, 0.4, 0.5, 0.6, 0.7, 0.8, or 0.9. Formally, the sDRW with better connectivity is defined as

$$W_{t+1} = (1-r)(M)\left(\frac{N_1 + N_2}{2}\right) + rW_t \qquad (2)$$

where, W_t is a vector of *i* node which is transmitted from *i*-1 node while *M* is an adjacency matrix developed from the original directed graph (with edges) to more strongly connected directed graph. As we stated in the previous section, weight plays an important role in nodes connectivity. Hence, weight of two connected nodes, N_1 and N_2 is implemented into the equation.

If the nodes have a strong connectivity towards previous nodes, then the vector from previous node towards it will be higher.

5. Direct Proof with Gene Expression Data

In this section, we present a direct proof of Significant Directed Random Walk that have better connectivity towards vertex compare to Directed Random Walk. By applying pathway data as references data and gene expression data as input data, we can obtain result of the equation. Hence, the ability of significant random walk can be proven.

Weight of the nodes are obtained from gene dataset, GSE19188 (Non-Small Cell Lung Cancer) [10], which will be processed in data pre-processing stage before implement into equation. Table 1 shows the weight of nodes after data pre-processing using Gene Chip Robust Multiarray Averaging (GCRMA).

Assume G; $V = \{1,2,3,4,5\}$ where G represent directed graph, while *V* represent vertex. Figure 1 shows part of the biological pathway, leukocyte transendothelial migration where highlighted nodes will be used in calculation and will be simplified into Figure 2.

Table 1. Weight of each node that implement in graph, G

Nodes	1	2	3	4	5	6
Weight	2.338914	8.47301	6.1441	3.102989	11.38365	5.149393

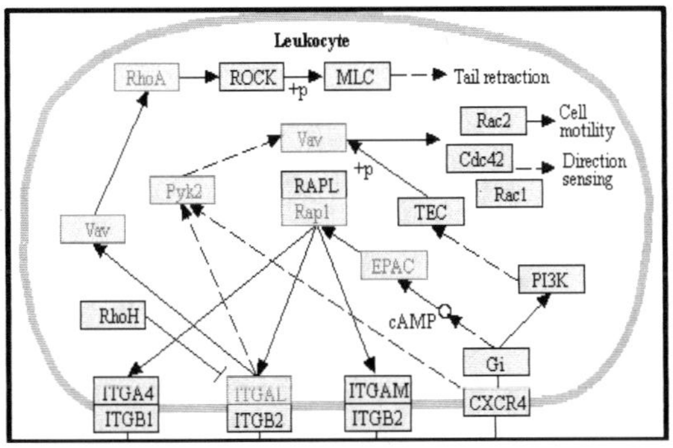

Figure 1. Gene sets that will be focusing on (highlighted)

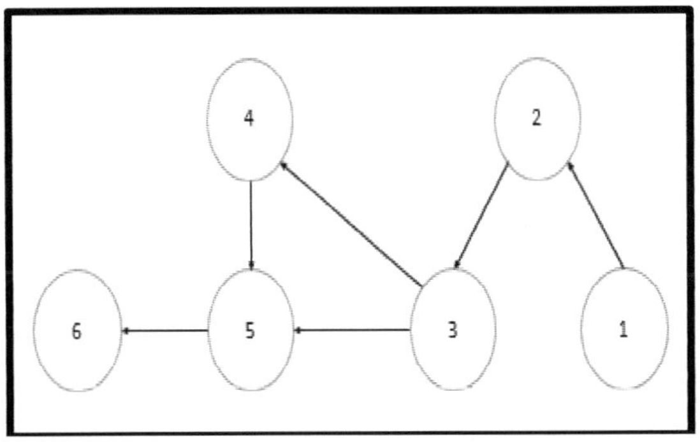

Figure 2. Graph G with nodes and edges after simplified from Figure 1

Figure 2 shows the nodes and edges that will be used in the calculation. The direction of the pathway is stated as below:
1 -> 2 -> 3 -> 4 ->5 > 6

Firstly, we will calculate the vector of significant directed random walk, followed by directed random walk, and compare the results in the end. The comparison of the results will be shown in Table 2. Initial vector, W_0 of first nodes (1) is zero because it is an initial node. Directed random walk will be proven by implementing the same data for comparison purposes. For first vector, W_t is set as 1.

Significant Directed Random Walk

$W_0 = 0$
$W_1 = (1\text{-}0.4)(1)((2.338914+8.47301)/2) + 0.4(0)$
$= 3.243577$
$W_2 = (1\text{-}0.4)(1)((8.47301+6.1441)/2) + 0.4(3.243577)$
$= 4.385133 + 1.297431$
$= 5.682564$
$W_3 = (1\text{-}0.4)(1)((6.1441+3.102989)/2) + 0.4(5.682564)$
$= 2.774127 + 2.273026$
$= 5.047153$
$W_4 = (1\text{-}0.4)(1)((3.102989+11.38365)/2) + 0.4(5.047153)$
$= 4.345992 + 2.018861$
$= 6.364853$
$W_5 = (1\text{-}0.4)(1)((11.38365+5.149393)/2) + 0.4(6.364853)$
$= 4.959913 + 2.545941$
$= 7.505854$

Directed Random Walk

$W_0 = 0$
$W_1 = (1\text{-}0.4)(1)(1) + 0.4(0)$
$= 0.6$
$W_2 = (1\text{-}0.4)(1)(0.6) + 0.4(0)$
$= 0.36$
$W_3 = (1\text{-}0.4)(1)(0.36) + 0.4(0)$
$= 0.216$
$W_4 = (1\text{-}0.4)(1)(0.216) + 0.4(0)$
$= 0.1296$
$W_5 = (1\text{-}0.4)(1)(0.1296) + 0.4(0)$
$= 0.07776$

Table 2. Comparison result of vector from node 1 to node 6

Vector, W	Significant Directed Random Walk	Directed Random Walk
W_0	0	0
W_1	3.243577	0.6
W_2	5.682564	0.36
W_3	5.047153	0.216
W_4	6.364853	0.1296
W_5	7.505854	0.07776

From the results above, we know the fluctuation of the vector in significant directed random walk is because of the weight. Weight plays an important role to attract the other nodes while the constant decline of vector in directed random walk is because of the previous vector. The connectivity (vector) will become weaker and weaker.

On the other hand, the connectivity between 2 nodes can also be proven by significant directed random walk. Hence, another calculation by using DRW and sDRW will be run to test the connectivity between first node and the other node. The paths are stated as below:
1 -> 2, 1 -> 3, 1 -> 4, 1 -> 5, 1 -> 6

The calculations by using sDRW and DRW are shown as below:

Significant Directed Random Walk

$$W_{1\to2} = (1\text{-}0.4)(1)\left(\frac{2.338914+8.47301}{2}\right) + 0.4(0) = 3.243577$$

$$W_{1\to3} = (1\text{-}0.4)(1)\left(\frac{2.338914+6.1441}{2}\right) + 0.4(0) = 2.544904$$

$$W_{1\to4} = (1\text{-}0.4)(1)\left(\frac{2.338914+3.102989}{2}\right) + 0.4(0) = 1.632571$$

$$W_{1\to5} = (1\text{-}0.4)(1)\left(\frac{2.338914+11.38365}{2}\right) + 0.4(0) = 4.116769$$

$$W_{1\to6} = (1\text{-}0.4)(1)\left(\frac{2.338914+5.149393}{2}\right) + 0.4(0) = 2.246492$$

Directed Random Walk

$$W_{1\to2} = (1\text{-}0.4)(1)(1) + 0.4\,(0) = 0.6$$

$$W_{1\to3} = (1\text{-}0.4)(1)(1) + 0.4\,(0) = 0.6$$

$$W_{1\to4} = (1\text{-}0.4)(1)(1) + 0.4\,(0) = 0.6$$

$$W_{1\to5} = (1\text{-}0.4)(1)(1) + 0.4\,(0) = 0.6$$

$$W_{1\to6} = (1\text{-}0.4)(1)(1) + 0.4\,(0) = 0.6$$

Table 3. Vector from node 1 to the other nodes

Vector, W	Significant Directed Random Walk	Directed Random Walk
$W_{1\to2}$	3.243577	0.6
$W_{1\to3}$	2.544904	0.6
$W_{1\to4}$	1.632571	0.6
$W_{1\to5}$	4.116769	0.6
$W_{1\to6}$	2.246492	0.6

From Table 3, with sDRW, we figure out the connectivity between node 1 and node 5 are strongest among the other nodes, while the weakest connectivity is between node 1 and node 4. While with DRW, the connectivity is remaining the same because the initial vector and first vector play the roles in determine the next vector. In sDRW, the reason to have such significant different of vector is because of the weight between each node.

6. Conclusion

Weight plays an important role between vertex and edge. Vector can be enhanced by implementing weight as one of the parameter. The equation shown is significant directed random walk. With the proven results after implementation, we believe that connectivity of nodes can be enhances with significant directed random walk as well as improve the vector. By comparing two different random walk, we figure out that the connectivity between nodes can be determine via vector. By using vector, the direction according to the pathway are fixed and possible to be simplify. With this, the pathway data can be used as references data to enhances the accuracy of the cancerous classification. sDRW proved that enhanced pathway can increases the accessibility of nodes towards the other significant nodes. With the enhanced pathway data, the result of accuracy can be increases due to fully utilized pathway data as references data. Fully utilized references data can help in increases the accuracy of cancerous classification.

Acknowledgments
We would like to thank Universiti Tun Hussein Onn Malaysia for supporting this research under the FRGS research grants (Vot numbers: 1559), also, thanks to Gates IT Solution Sdn Bhd for the whole support.

References
[1] Woolard, E. W., Einstein, A., Furth, R., & Cowper, A. D. (1928). Investigations on the Theory of the Brownian Movement. The American Mathematical Monthly, 35(6), 318. DOI:10.2307/2298685
[2] Pearson, K. (1905). "The Problem of the Random Walk". Nature. 72 (1865): 294. DOI:10.1038/072294b0
[3] Edward A Codling, Michael J Plank, Simon Benhamou. 2008. Random walk models in biology. J. R. Soc. Interface. (August 2008), 813-834. DOI= 10.1098/rsif.2008.0014.
[4] Bapat, R., Kalita, D., & Pati, S. (2012). On weighted directed graphs. Linear Algebra and its Applications, 436(1), 99-111. DOI:10.1016/j.laa.2011.06.035
[5] Aldous and Fill, Reversible Markov Chains and Random Walks on Graphs. Retrieved February 15, 2017, from http://www.stat.berkeley.edu/~aldous/RWG/book.html.
[6] Liu, W., Li, C., Xu, Y., Yang, H., Yao, Q., Han, J., . . . Li, X. (2013). Topologically inferring risk-active pathways toward precise cancer classification by directed random walk. Bioinformatics, 29(17), 2169-2177. DOI:10.1093/bioinformatics/btt37
[7] Bender, M. A., & Ron, D. (2002). Testing properties of directed graphs: acyclicity and connectivity. Random Structures and Algorithms, 20(2), 184-205. DOI:10.1002/rsa.10023.abs
[8] Cai H, Ruan P, Ng M, Akutsu T. Feature weight estimation for gene selection: a local hyperlinear learning approach. BMC Bioinformatics. 2014;15:70. doi: 10.1186/1471-2105-15-70.
[9] Seah, C. S., Kasim, S., and Mohamad, M. S.,"Specific Tuning Parameter for Directed Random Walk Algorithm Cancer Classification," International Journal on Advanced Science, Engineering and Information Technology, vol. 7, no. 1, 2017. [Online]. Available: http://dx.doi.org/10.18517/ijaseit.7.1.1588.
[10] Hou J, Aerts J, den Hamer B, van Ijcken W et al. Gene expression-based classification of non-small cell lung carcinomas and survival prediction. PLoS One 2010 Apr 22;5(4):e10312. PMID: 20421987

IOP Conf. Series: Materials Science and Engineering **235** (2017) 012005 doi:10.1088/1757-899X/235/1/012005

An Intelligent Agent-Controlled and Robot-Based Disassembly Assistant

Jan Jungbluth[1,2,3,4], Wolfgang Gerke[2,5] and Peter Plapper[3,6]

[1] Technology Research Group, SEW EURODRIVE, Bruchsal, Germany
[2] Robotic and Control Group, Trier University of Applied Sciences (TUAS), Environmental Campus Birkenfeld, Birkenfeld, Germany
[3] Faculty of Science, Technology and Communication, University of Luxembourg, Luxembourg, Luxembourg

{[4]jan.jungbluth | [5]w.gerke}@umwelt-campus.de, [6]peter.plapper@uni.lu

Abstract. One key for successful and fluent human-robot-collaboration in disassembly processes is equipping the robot system with higher autonomy and intelligence. In this paper, we present an informed software agent that controls the robot behavior to form an intelligent robot assistant for disassembly purposes. While the disassembly process first depends on the product structure, we inform the agent using a generic approach through product models. The product model is then transformed to a directed graph and used to build, share and define a coarse disassembly plan. To refine the workflow, we formulate "the problem of loosening a connection and the distribution of the work" as a search problem. The created detailed plan consists of a sequence of actions that are used to call, parametrize and execute robot programs for the fulfillment of the assistance. The aim of this research is to equip robot systems with knowledge and skills to allow them to be autonomous in the performance of their assistance to finally improve the ergonomics of disassembly workstations.

1. Introduction

Disassembling is involved in many processes, for example remanufacturing, corrective maintenance, proper disposal and manufacturing. However, fully automated disassembly lines are, compared to product assembly lines, rare. One reason is that disassembling at the end of a product lifetime is much harder to automatize than assembling. In disassembly we have to cope with fouling, wear, damaged or only absent parts. Furthermore, we have to deal with product manipulations, such as individual extensions or improvised fixes, which are not obviously visible. It is the unpredictable condition of a product that prohibits further automation. Even if better sensor technology could identify inappropriate product conditions, it would be impossible, or highly expensive, to treat all possibilities in a fully automated manner. Other challenges are the small lot sizes or individualized products (lot size 1). Today's fully automated processes are not flexible enough for the treatment of different products or variants. In corrective maintenance, it is a normal case to have an unsteady, unpredictable flow of different products. Dismantling processes in corrective maintenance may also have different target stages, depending on what part has to be replaced. Furthermore, there is, especially in central waste recycling plants, a lack of information about the product structure. For example, types of materials in the product. These are reasons why disassembly workplaces stayed unautomated, thus resulting in a bad situation for the workforce, who remain exposed to health problems due to heavy workload. In addition, the economic fitness of disassembly processes is highly reduced by the substantial amount of manual labor in the destructive disassembly process. Consequently, important concepts for environmental protection, such as remanufacturing, cannot spread wider into industry.

A solution for this problem would be intelligent robot-based assistants, which would be a compromise between automation and manual labor with great advantages in the disassembly domain. One advantage is that only humans have the cognitive abilities to identify and handle the aforementioned unexpected situations. Thus, humans are able to ensure the overall success of the process by contributing their awareness to adapting the process to a situation [1]. The robot instead can provide assistance with power and endurance over the complete disassembly process and thereby improve the ergonomics of the workplace. Therefore, multi-skilled robots may take over automatable tasks, such as unscrewing, or support the execution of tasks, e.g., handling of heavy parts. Especially over the 10 last years, with the development of lightweight and force-sensitive robots, new robot skills have extended the robots' fields of application. Those new skills provide realizable "assistance opportunities" in disassembly procedures, and some of them could be integrated into one robotic system. However, once integrated, the problem arises of controlling the multi-skilled robot and its behavior to assist the user in a situation- and goal-oriented manner. Even providing the system with needed precise positions information is big issue [2]. Manually programming the whole process would not be meaningful [3], especially if we consider many different products. Only intelligent systems with higher autonomy could support and interact fluently with humans in such a dynamic environment [4, 5]. Therefore, the system needs to identify the disassembly process and the work contend by itself. Then, the robot must know how to assist the user in a certain task; it also needs necessary information and skills to perform this task. To meet these challenges, we take the approach of an informed software agent, the architecture of which we present and explain in Section 2. In Section 3, we explain the kind of information that is provided to the agent and in section 4 we clarify how we decompose the complex disassembly problem into individual subproblems. Coarse disassembly planning is used to identify the task sequence that strips down the assembly. In Section 5, we present our approach to assigning the work content of each task to the participants through a planning search in the possible situation space. Explained in Section 6 is how we execute the determined detailed plan by invoking and parametrizing robot programs to finally control the robot assistance behavior.

2. The Agent-Based Robot Control Architecture

Today's industrial robot controllers are not suitable for human -robot interaction in complex environments such as disassembly workstations [1]. In such environments, the objective and the boundary condition change from task to task and require some planning, higher skills, and knowledge to succeed efficiently. Our approach to empower industrial robots to this application field is to overlay the robot controller with an intelligent agent-based control system, which does high-level planning, defines and controls the robot behavior, e.g., the type of assistance, and guarantees the necessary information to the underlying robot controller. The developed software agent, which we present in this section, consists of different software modules (see figure 1). In the agent's "Knowledge Base" module, we store a symbolic representation of the product structure, which we call the product model. The model contains information about the parts and how the parts are connected with each other in the assembly. Therefore, we model different classes of parts and connections. We also model the actions that the actors (the user and the robot) provide to the overall system, to act in the environment or on the product structure. We explain all the stored information in further detail at the time of its use, following the systems information flow and processing. In the "Coarse Planning" module we create, depending on the disassembly objective and the product model, a coarse plan that consists of a sequence of disassembly steps. Each disassembly step consists of one or more tasks. A task defines an independent subproblem, which is the removal of a specified connection and the referenced parts. Moreover, we use the coarse planning to assign the means of production, such as tools, robot effectors, and carriage cases, to a task. This is done through the "Stock Manager" module. Depending on the connection type, different assistance behaviors of the robot are available and selectable in a task by the user. Furthermore, the user can rearrange the task order and manipulate parts' condition states, for example, to mark a damaged part, over the human-machine-interface (HMI) module to adapt the process to the circumstances and his/her will. The coarse plan is then processed task by task by the "Detail Planning" module, in which we first create a discrete state space representation of the current and the target situations. A search through the possible state space, which considers the actions provided by the user and the robot, leads to a set of possible action sequences from which the fittest sequence is selected for execution. The selected action sequence is then performed, monitored and synchronized by the "Control" module through advising the user over the "HMI" module or invoking and parameterizing programs on the underlying robot controller.

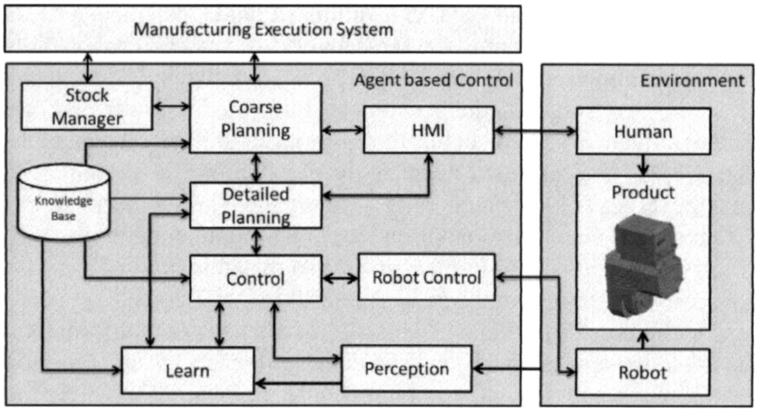

Figure 1. The agent's architecture and the information flow between the different software modules.

Currently under construction is the "Perception" module. In the first place, it is foreseen to track the human hands to synchronize the process without action-executed confirmation-button pressing. We use the "Perception" module also to recognize and interpret gestures and voice commands to alter the robot behavior in the execution phase. The "Learn" module is considered in the agent architecture to adapt the assistance behavior to an individual user. A possible purpose of machine learning is to predict the user's desire for an assistance behavior by comparing the current situation with similar situations from recorded older interactions. Also, the agent could improve the coarse planning, particularly the ordering of parallel executable tasks, by learning from manually adapted plans. Furthermore, recording and analyzing such disassembly processes could produce valuable data and lead to deeper insights into the disassembly process. Next, we explain the system in more detail, starting in the knowledge base and considering the product model.

3. The Product Model

A technical product is an assembly of parts that are linked together through connections. The step-by-step removal of these connections is the process of disassembling. The sequence of removing the connections is not arbitrary but partially defined through the product structure. Finding this disassembly sequence is known as disassembly planning, which is the topic of Section 4. So to generate the disassembly sequence later, we first have to present the structure in an appropriate and machine readable manner. Disassembly planning is an ongoing research field which also gains interest, through the assembly-by-disassembly approach, from the well-studied area of assembly planning. The majority of reviewed publications in this area use undirected and directed graphs as well as hypergraphs to represent the product structure [6–10]. Other approaches are based on petri nets [11], description logic and object-oriented models [12], and more recent approaches are based on ontologies [13–16]. Most works focus on sequence generation and thus lack in providing metadata for the human-robot interaction. For our usage, we have decided to develop an easy-to-use, light but comprehensive model to represent the product structure. After careful consideration, we have chosen the object-orientated approach based on its advantages of information encapsulation, easy implementation, and

IOP Conf. Series: Materials Science and Engineering **235** (2017) 012005 doi:10.1088/1757-899X/235/1/012005

extension. In addition, we can create other structures from the object-orientated model. We now explain the two fundamental classes of which our product model is composed.

3.1. *Part Classes*

In contrast to the most works which are based on [7], we do not tie functional parts to connections; instead, we clearly separate parts and connections into our fundamental classes. To distinguish between different classes of parts that have different needs of information in the sense of disassembly, we use a flat taxonomy-like hierarchy of classes (see figure 2). Each class inherited from the super class carries common and more specialized information in its attributes. A common attribute, for example, is the position and orientation of a part in reference to the main coordinate system of the assembly. More specialized information could be the thread length or the driver style of a screw class instance. An overview of the attributes we use in the super class is listed in the table of figure 2. Whenever a component indicates some special needs of information in the disassembly process, we can create a new class or extend a similar one. Chiefly, this is necessary for fasteners or connection techniques, but it could also be used to model fluids or gases. While the removal processes of a joint can have parameters that depend on the attributes of a part, we uncouple this information from the process. For example, the thread length and the thread pitch of a screw define the number of twists until the screw is loose. In this sense, it is a great advantage to use variable process parameters and to uncouple part-dependent facts from processes. Furthermore, we implemented state descriptions and qualifiers for each part class to represent each class's possible and current condition. This could be used to mark a part that it is damaged or missing. Beyond representing several parts, it is essential to model subassemblies. Mostly, it is not advisable to strip down a subassembly in another assembly. Typically, it is better to fractionalize subassemblies after their removal from the main assembly. So treating a subassembly as a special part class is beneficial in the manner that subassemblies gets completely removed from the main assembly and then further dismantled.

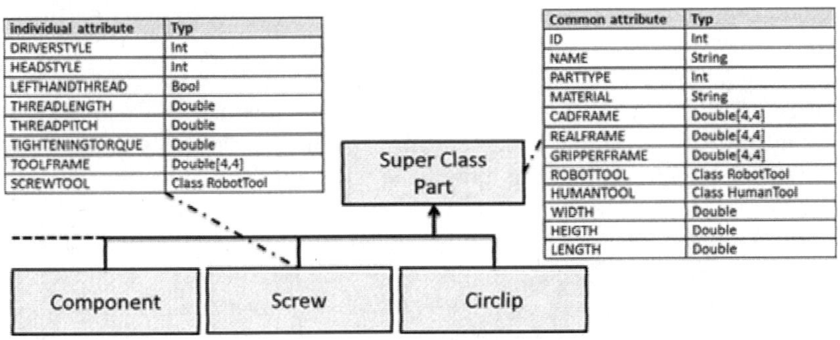

individual attribute	Typ
DRIVERSTYLE	Int
HEADSTYLE	Int
LEFTHANDTHREAD	Bool
THREADLENGTH	Double
THREADPITCH	Double
TIGHTENINGTORQUE	Double
TOOLFRAME	Double[4,4]
SCREWTOOL	Class RobotTool

Common attribute	Typ
ID	Int
NAME	String
PARTTYPE	Int
MATERIAL	String
CADFRAME	Double[4,4]
REALFRAME	Double[4,4]
GRIPPERFRAME	Double[4,4]
ROBOTTOOL	Class RobotTool
HUMANTOOL	Class HumanTool
WIDTH	Double
HEIGTH	Double
LENGTH	Double

Figure 2. The inheritance hierarchy of part classes with details of common and individual attributes.

Also considered is information for the further treatment of the parts. In a remanufacturing process, we could be interested in adding information to the parts about a test and a rework process a part has to pass to get back into production. For appropriate recycling, the material is an important attribute. This approach to equip the robot with the necessary information looks plausible, but it also means a substantial amount of knowledge engineering. Some of this information is already stored in today's CAD systems in a form of standard part libraries and could be used as a source of information for the product data model. Some information needs to be consolidated and made available in a clearly structured manner that is readable for humans and machines.

3.2. *Connection Classes*

Similarly to the part classes, we build hierarchies with connection classes (see figure 3). Some common attributes are also illustrated in the table of figure 3. A connection defines what kind of liaison is between two or among more parts and could be of any type. From a face-to-face contact, a weld joint or a magnetic attraction, any kind of liaison can be designed and implemented. The real power of the different connection classes is that they describe a formal process. If we can describe a certain state and objects in such a process, then we have a situation. Knowing the situation gives the assistant the ability to link commands dynamically with context information. In Section 5, we describe a formal process definition that makes use of a state representation and a set of actions, which could be performed by the acting agents (e.g., the user and the robot). We use this formal process description in detailed planning (Section 5) to find the best sequence of actions to remove the bound. As we see more clearly in Section 4.1, we can represent the product structure in a symbolical (machine-readable) manner by describing the parts and connections of an assembly.

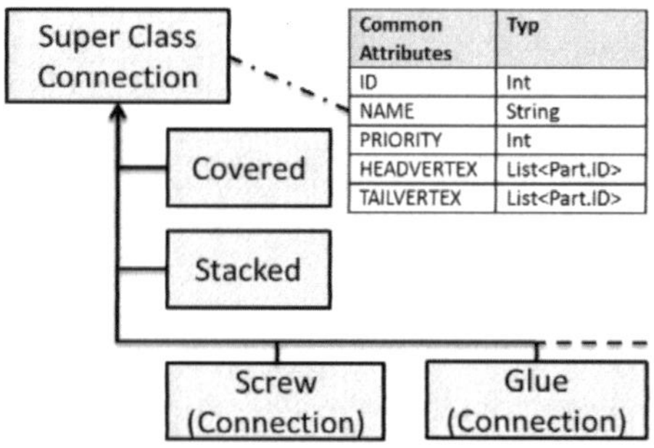

Figure 3. Inheritance hierarchy of the connection class. The ordering implied the disassembly priority of each class from low to high. Also illustrated are the common attributes of the super class.

For the efficient manual design of these connections, we integrated functions into the NX 10 CAD system, which also generates the instances of the part classes during the connection creation. In future work, the product model will be automatically created by the product configurator. The format used to store the part and connection instances for a product model will be based on the XML format and be transferred via RFID directly from the assembly or via web services to the software agent.

4. Coarse Disassembly Planning

Coarse disassembly planning reduces the amount of process definition done by the user to a significant level. While the disassembly task can vary among corrective maintenance, remanufacturing and recycling, we have to be able to create substructures of the complete disassembly for the exchange of wear and tear parts and target-oriented plans for the removal of valuable components. We also have to generate plans for the full dismantling of the product for recycling purposes. To achieve this, the disassembly task is communicated to the agent through the product model and the part to be removed. In this form, we are able to determine a plan for the removal of valuable or broken parts and, through selecting the root component of the assembly, the complete disassembly of a product. In the reviewed works [6–16] on assembly or disassembly planning, the main focus was on finding a sequence in which the parts could be added or removed to create a complete assembly or disassembly. It was not considered, to build substructures of assemblies, or to do further task and process planning. The used techniques vary depending on how the product structure was modeled. Most techniques used on graphs are mathematical and based on the adjacency matrix [6-9]. The inference is used by ontologies and description logic approaches [13–16], and rule-based systems use forward or backward chaining. Also found in the reviewed papers are applications of fuzzy logic and genetic algorithms [11]. We have decided to use the simple but powerful approach of topological sorting and do further task and process planning. We build the coarse plan in five steps. The first step is to create the product graph from the product model. In the second step, we create the minimal graph with the selected part as "root". Then we create a sequence of disassembly steps and tasks through topological sorting in the third step. In the fourth step, we assign the means of production to each task. In the last step, the user has to define the type of assistance we want in a certain task. Each step is shortly discussed as follows:

4.1. Creating the Product Graph

IOP Conf. Series: Materials Science and Engineering **235** (2017) 012005 doi:10.1088/1757-899X/235/1/012005

We represented the product structure internally through a directed graph **G**(**V**, **E**), which consists of a set of vertices **V** and edges **E**. A vertex **v** represents a part instance, and an edge $\mathbf{e}(\mathbf{v}_i, \mathbf{v}_j)$ represents a link between two parts with the direction from the head $\mathbf{v}_i$ to the tail-vertex $\mathbf{v}_j$.

A connection instance may have several edges depending on the connection type to better depict the interactions between parts. The graph is automatically created by the parts and connections described in the product model. For a better understanding of this topic, we will explain the graph creation and the part and connection models from the previous topic. For further explanation, kindly compare the sectional view of an electrical drive (in figure 4) and its product graph (in figure 5).

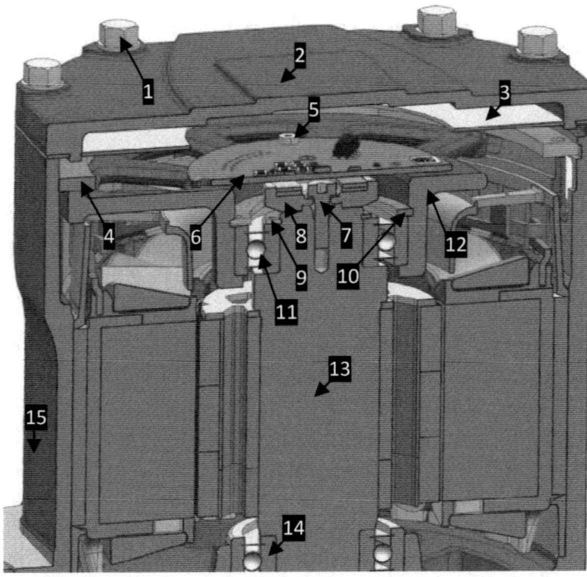

Figure 4. A cross-sectional view of an electrical. drive with numbered parts.

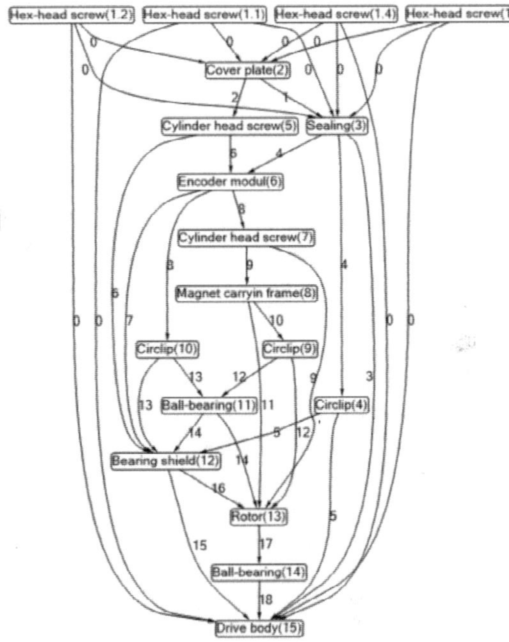

Figure 5. A product graph of the electrical drive of figure 4. The numbers on the edges are the IDs of connection instances.

We now explain three different connection types, two part types, and how we represent them in the graph. First, we consider the bolted joint of the four hex-head screws (Number 1.1–1.4 in figure 5) that link together the cover plate (2), the sealing (3) and the drive body (15). The four screws are all instances of the "screw" part class. The sealing, the cover plate, and the drive body are instances of the "component" part class.

The bolted joint is symbolically described by only one instance from the "screw" connection class because the screws are identical, and they link the same parts. In the screw connection instance, we define the screws as head- vertices and the other components as tail-vertices.

This connection is represented in the graph with sixteen edges, from each screw to any of the other three connected parts. The next connection we have to specify is the cover plate, which is lying on the sealing. We represent it by an instance of the "stacked" connection class. This connection is represented by one edge from the cover plate to the sealing. An instance of the "covered" connection is used to describe the situation that one part avoids access to another part without any physical interaction. This case applies to the cover plate and the cylinder head screw (5). The covered instance is also illustrated on the graph by one edge from the cover plate to the screw.

4.2. Disassembly Planning

The part to be disassembled defines the start node of a breadth-first algorithm, which adds all parent nodes and corresponding edges to a new subgraph. Through topological sorting of the subgraph, we then create the disassembly sequence. In each step of the sequence, we identify part nodes that have no ingoing edges and could be removed. Through the outgoing edges of each part node, we get related connection instances. Then, a new task is created with the connection and part instances thereby taking care that parts that belong to the same connection instance are grouped together in the same task (in the current disassembly step). Moreover, if a part is attached to more than one connection, then we sort the connections by their disassembly priority in the task. The disassembly priority describes which connection of a set of liaisons has to threaten first. For example, if part A lies on part B and between A and B is a glue, then it is better to warm up the parts until the glue loses its strength and then remove part A. Because multiple parts and connections could be removed independently in a disassembly step, we can have multiple parallel executable tasks. If all removable parts are handled in one disassembly step, we delete the nodes and edges and create a new step. This process repeats until the subgraph has no more nodes.

4.3. Assigning the Means of Production

Especially in remanufacturing processes, we have to separate unequal, and usable from unusable, parts for their further treatment. To do so, we have to store the parts in different transport boxes and keep track of which part is in which box. Assigning the right box to a part, and vice versa, is one duty of the local "stock manager" module. The stock manager uses the width, height and length attributes of a part to find a suitable box in the local stock. The box is then assigned to the task and part. If the box is not available in the local stock, then the stock manager produces an order. While we model all process-involved objects in our environment, we have a class that represents boxes. A box object has attributes, such as its storage dimensions or their position and orientation in reference to the robot work frame, and qualifiers to represent their current state, e.g., if there is a box available, or it is ordered. Furthermore, the local stock manager determines the needed human and robot tools by accessing the attributes of a part with respect to the connection type. For example, a screw instance has three different tools: the tool for human use, such as a screwdriver, a robot tool for manipulating the part, usually a gripper, and a robot screw tool to loosen the joint. Like the box object, each tool is represented by an instance of a tool or robot tool class and has attributes, such as their positions, and state qualifiers. Furthermore, we assign the actors, e.g., the user and the robot, to the tasks to have all process relevant entities together.

4.4. Plan Manipulation and Assistance Definition

The sequence generated through the agent is a partially ordered plan. So there are different possible processes and the created one does not have to match a user's expectation. Furthermore, there could be other reasons that the user wants to rearrange or manipulate the found order. To rearrange the process, the user can move tasks back and forth in the disassembly sequence and manipulate the process in other methods by modifying the state qualifiers of the parts and objects to adapt the process to the current product state. More input is needed regarding the fact that the user has to define how he expects to be supported. Therefore, the user has to choose for each connection one type of assistance from a set of recommended assistance behaviors. At the moment, this is done by right clicking on the link with the computer mouse, which is a very unnatural method to communicate and flow-break. Using multimodal communication, through gestures and voice commands, to define or change the form of assistance would be a great advantage and is currently under investigation. To reduce the amount of assignment in this kind, future work will have to investigate if machine learning or case-based reasoning is able to predict a user's wish for assistance with tolerable accuracy.

To summarize the steps to this point, we have used a decomposition strategy [17] to split the complex disassembly problem into smaller, independent subproblems. The subproblem we consider is the removal of a certain type of connection, which is a process that we can generally describe formally and solve efficiently because of its smaller size. Furthermore, we can think of the coarse planning in a manner that it determines work that has to be done, without assigning it to the user or robot. The detailed planning in the next step solves the subproblem and assigns the work to the actors.

5. The Detailed Disassembly Planning

In the ideal case, the manual removal of a connection takes place through a fixed sequence of actions. In disassembly workstations, we regularly find cases in which the sequence has to be adapted, probably with other actions, to produce the wanted output. Collaborative work also means that actions might be executed by one or another agent, in a dynamic manner that produces many variants of the process. The actors can also provide assistance to each other, for example, the user can change a robot effector so that there are even more possibilities. With this combinatory problem in mind, we have decided that the process should not be explicitly defined through the use of fixed finite state machines or petri nets. Instead, we use a search approach to find a suitable sequence of actions to solve our subproblem. To represent our issue as a search problem, we have first to decide on a vocabulary of conditions, objects, and actions. Then, we have to encode actions from our domain and define a problem instance by defining the initial and the target conditions. We now explain how we generate the initial state and target state, represent the actions and what algorithms we used to search for a solution by using an easy example task. We consider the loosening of a "stacked" connection, in which part A is simply lying on part B.

5.1. The Task State Description

The state we want to represent depends on the connection type and the involved objects, such as part(s), tool(s), boxes and the acting agents. Since we have already collected all these objects in a task description, we can create the initial state by merging all objects' state descriptions. Some objects' state values are predefined, for example, the *PartState*, *PartPos*, and *ConnectionState* (see Table 1), but could be manipulated by the user. The value of the *BoxState* and *BoxPos* are defined by the stock manager, depending on whether, if the box is ordered, it is full or ready to use. The robot states are defined by sending queries to the robot controller. The users *HandPos* state will be tracked, and the *HandState* value is estimated by the perception module. These state values represent literals, which formally represent the condition of the task at the beginning, the end, and in between through a state vector. In a state, we also save the parent's ID, the costs and the action type that created the state. For the initial state, these attributes are zero. The end condition of the task state is partially defined, through the default values of the *ConnectionState* and *PartPos* (see Table 1). It is the users' choice to add other state dimensions and values to the end condition to define it more precisely.

Table 1. Observed objects in the task. Each object can have several dimensions and state values. The underlined values indicate predefined states of the initial task state. The literals in cursive are default values for the target state of the task.

Object type	State dimension	State values
Stacked Connection	Connection-State	isStacked, isDetached, *isRemoved*
Human	HandState	isEmpty, isNotEmpty
	HandPos	atUnknownPos, atPartPos, atBoxMagazinPos, atRobotGuidePos
Robot	RobotState	isUnknown, isIdle, isRunning, isGuided
	RobotPos	atUnknownPos, atHomePos, atPartPos, atBoxMagazinPos, atRobotToolMagazinPos,
RobotTool	RobotToolState	isUnknown, isOpen, isClosed
	RobotToolPos	atUnknownPos, atRobotFlange, atRobotToolMagazinPos
Part	PartState	isOK, isNOK,
	PartPos	atUnkownPos, atPartPos, atGripper, atHand, *atBox*
Box	BoxState	isAvailable, isOrdered, isFull
	BoxPos	atUnknownPos, atBoxMagazinPos

5.2. Describing actions

Methods to describe or analyze manual and robot-automated tasks are well known, for example, Method-Time-Measurement (MTM) and Robot-Time-And-Motion (RTM). For the new type of collaborative work, there is quite a lack of methods. An adapted Method-Time-Measurement as process logic for cognitive automated assembly was mentioned in [18], which did not mention a human-robot interaction. Other studies such as [19] only link the basic Methods-Time-Measurement (MTM-1) system to equivalent robot actions without any collaborative actions. We have decided to describe actions only with the motion and end-effector elements of RTM, thus treating the human as a robot, and added a new element: collaborative actions. The removal of a stacked connection is described by the actions listed in Table 2. Each action has a precondition and transmission vector, which describes when the action is executable and how it affects the state. Each action defines which program on the robot controller is called when the action gets executed. The attribute list of an action is used to parametrize the robot program with the needed parameters. For example, the actions *GotoPartPos* and *GotoBoxMagazinPos* in Table 2 have the position of the part and box in the actions parameter list. Both actions call the same robot program, which only moves from its current position to the "send" part or box position. Furthermore, actions have a cost value that is dynamically assigned with respect to the user's choice of assistance. If an action is part of the assistance the user wants, it gets a lower cost assigned. In this example, we have mentioned three different robot behaviors. The *Manual* behavior defines only human actions and, therefore, manual work. *Automatic* means that the robot works autonomously. The *Collaborative* form describes that the user guides the robot to the unknown part position, and the robot can then grasp the part and put it in the transport box. We use this approach to provide to the users a set of known and predictable robot behaviors that stay adaptive to allow necessary modifications.

Table 2. All actions that could take place in the process of removing the stacked connection. The different assistance opportunities and the corresponding weighting of the action are indicated by L for low and H for high action costs.

Actor	Type	Action	Manual	Automatic	Collaborative
Human	Motion Element	GotoPartPos	L	H	H
		GotoBoxMagazinPos	L	H	H
		GotoRobotGuidePos	H	H	L
	End-effector Element	GrabPart	L	H	H
		ReleasePart	L	H	H
	Collaborative Element	GuideRobotToPartPos	H	H	L
Robot	Motion Element	GotoHomePos	H	H	H
		GotoPartPos	H	L	H
		GotoBoxMagazinPos	H	L	L
	End-effector Element	OpenGripper	H	H	H
		CloseGripper	H	H	H
		GrabPart	H	L	L
		ReleasePart	H	L	L
	Collaborative Element	GetGuidedToPartPos	H	H	L

5.3. Searching for solutions

Although the problem of removing part A, which is lying on part B, does not look very complicated, we have to be aware of the combinative size of the problem. In a naive breadth-first search approach, with a branching factor of 8 and the minimal process depth of 4, we have 8^4 (4,096) different paths to compute. A method of improving the algorithm is the use of an extended state list that stops us from extending paths on nodes we have already extended. Furthermore, we have implemented a policy that forbids the algorithm to use two motion elements in a row. This already gives us (for the problem mentioned) a good computational state space. To find the best path of actions, we have used a branch and bound search with extended list and the policy (see figure 6). If the initial state could be a starting point of all assistance behaviors, we would get without dynamic weighting only one form of assistance. With dynamic weighting, we get the user selected form of assistance. This approach of assigning the work content to the agents might seem overloaded on the problem we mentioned, but it uses a generic

IOP Conf. Series: Materials Science and Engineering **235** (2017) 012005 doi:10.1088/1757-899X/235/1/012005

approach that we can use for all other disassembly subproblems, such as screwing and so on. Also, our technique gives the assistance system a highly adaptable behavior. If the user leaves the workplace, his/her actions are not usable, and the only behavior is the *Automatic* one. If the part position is unknown in the *Automatic* mode, the next found assistance form is the *Collaborative* behavior (see figure 7). In the case that the search algorithm might not find a suitable sequence of actions (solution), it can explain its reasoning process through the symbolic structure of the situation state space. Also, a great advantage of the search approach is that when we build these detail plans, task after task, we do not have to worry about plan merging, because at each task we create a new plan on the actual situation.

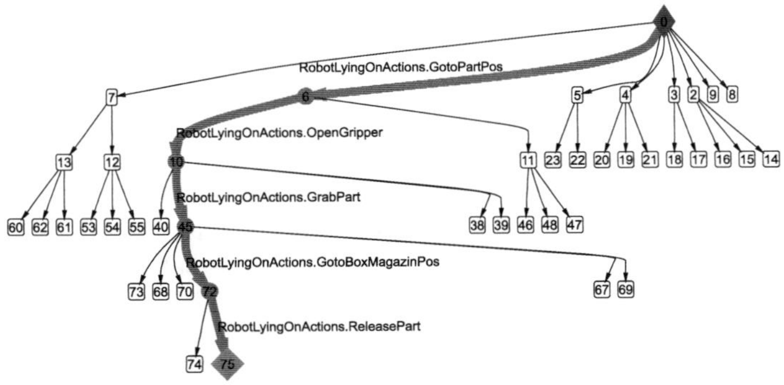

Figure 6. A state space created by the search algorithm. From the initial state (blue diamond) and with the *Automatic* assistance behavior, a solution state (red diamond) was found, and the robot actions on the green path could be executed.

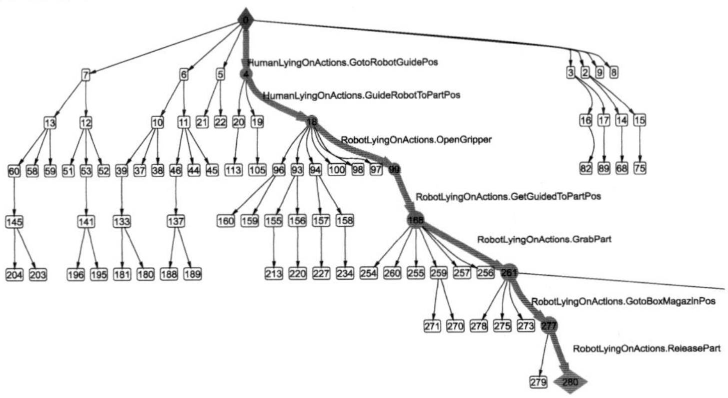

Figure 7. This state space was created by the search algorithm also with the *Automatic* assistance selected, but through the fact that the part position is unknown, the solution consists of the actions of the *Collaborative* assistance behavior.

6. Execution of the Detailed Plans

The "Control" module of the agent is responsible for the execution of the robot programs, for the guidance of the user and for the synchronization of both. The robot programs are invoked and started through a TCP/IP socket connection from the Robot Controller. Then, the robot program starts and connects as a client to our agent to receive the actions parameters list items. At the moment, the disassembly is processed task by task and action after action. The synchronization of the user and robot actions rely on the user's pressing of the confirmation button. To improve the synchronization, we want to track the human hand movements to estimate if an action was carried out. This lets us keep track of the user progress. For example, if we observe that the user moves his hand to the part position and then to the box position, we can assume that he has removed the part. This would be a very efficient way to synchronize the users' and robots' actions. Another important goal is to enable the parallel execution of tasks. The reduced robot motion in collaborative workplaces and the step-by-step workflow wastes much time, and the user gets bored waiting for his/her turn. So this is a very crucial skill for the assistant. Another ambition is to enable the user to give commands through a gesture or voice in full operation. This could be used to give simple commands, such as "stop," "open gripper" or "close gripper," or to change the assistance behavior completely and force detailed replanning.

7. Conclusion

In this paper, we have represented an approach of a robot-based disassembly assistant controlled by an informed software agent. We have discussed the need for a common workflow for a fluent, safe and purposeful assistance in collaborative disassembly. We have further described our approach to inform the agent with product models and our developed two-stage process of workflow planning. Therefore, we have explained the first planning step, which is based on the product graph, topological sorting, and task planning algorithms. We have also illustrated the second task-based planning step, which has focused on the refining of the workflow depending on the situation and the user-chosen assistance through a branch and bound search algorithm. In Section 6, we have discussed executing the robot assistant behavior. Finally, we can summarize that the information and methods we have provided to the robot assistance system enable higher autonomy to perform valuable assistance.

Acknowledgments

This research is conducted in a cooperation among the Company SEW EURODRIVE, the University of Luxembourg and the University of Applied Sciences Trier - Environmental Campus Birkenfeld.

References

[1] Wolfgang Gerke, *Technische Assistenzsysteme: Vom Industrieroboter zum Roboterassistenten*. Berlin: De Gruyter Oldenbourg, 2015.
[2] S. L. Kendal and M. Creen, *An Introduction to Knowledge Engineering*. London: Springer-Verlag London Limited, 2007.
[3] A. Perzylo *et al.*, "Intuitive instruction of industrial robots: Semantic process descriptions for small lot production," in *2016 IEEE/RSJ International Conference on Intelligent Robots and Systems (IROS)*, pp. 2293–2300.
[4] B. Ludwig, *Planbasierte Mensch-Maschine-Interaktion in multimodalen Assistenzsystemen*. Berlin: Springer Vieweg, 2015.
[5] R. R. Murphy, *Introduction to AI robotics*. New Delhi: Prentice-Hall of India, 2004, c2000.
[6] G. Dini, F. Failli, and M. Santochi, "A disassembly planning software system for the optimization of recycling processes," (en), *Production Planning & Control*, vol. 12, no. 1, pp. 2–12, 2001.
[7] L. S. Homem de Mello and A. C. Sanderson, "A correct and complete algorithm for the generation of mechanical assembly sequences," in *Proceedings, 1989 International Conference on Robotics and Automation: A correct and complete algorithm for the generation of mechanical assembly sequences*, May. 1989, pp. 56–61.
[8] L. S. Homem de Mello and A. C. Sanderson, "AND/OR graph representation of assembly plans," (en), *IEEE Trans. Robot. Automat.*, vol. 6, no. 2, pp. 188–199, 1990.
[9] Y. Jyh-Cheng and L. Yi-Ming, "The Structure Representation for the Concurrent Analysis of Product Assembly and Disassembly," Department of Mechanical and Automation Engineering, 2005.
[10] H. J. Kim, S. Kernbaum, and G. Seliger, "Emulation-based control of a disassembly system for LCD monitors," *The International Journal of Advanced Manufacturing Technology,* vol. 40, pp. 383–392, 2009
[11] G. D. GRADI, "PETRI-NET Modelling of an Assembly Process System," Institute of Information Theory and Automation - Department of Adaptive Systems, Prague, Czech Republic

[12] Xu, C. Wang, Z. Bi, and J. Yu, "Object-Oriented Templates for Automated Assembly Planning of Complex Products," (en), *IEEE Trans. Automat. Sci. Eng.*, vol. 11, no. 2, pp. 492–503, 2014.
[13] M. Merdan, W. Lepuschitz, T. Meurer, and M. Vincze, "Towards ontology-based automated disassembly systems," in *IECON 2010 - 36th Annual Conference of IEEE Industrial Electronics*, pp. 1392–1397.
[14] N. Lohse, H. Hirani, S. Ratchev, and M. Turitto, "An ontology for the definition and validation of assembly processes for evolvable assembly systems," in *(ISATP 2005). The 6th IEEE International Symposium on Assembly and Task Planning: From Nano to Macro Assembly and Manufacturing, 2005*, Jul. 2005, pp. 242–247.
[15] S. Chen, J. Yi, H. Jiang, and X. Zhu, "Ontology and CBR based automated decision-making method for the disassembly of mechanical products," (en), *Advanced Engineering Informatics*, vol. 30, no. 3, pp. 564–584, 2016.
[16] S. Balakirsky, Z. Kootbally, C. Schlenoff, T. Kramer, and S. Gupta, "An industrial robotic knowledge representation for kit building applications," in *2012 IEEE/RSJ International Conference on Intelligent Robots and Systems (IROS 2012)*, pp. 1365–1370.
[17] Q. Yang, *Intelligent planning: A decomposition and abstraction based approach*. Berlin: Springer, 1997.
[18] B. Britzke, *MTM in einer globalisierten Wirtschaft: Arbeitsprozesse systematisch gestalten und optimieren*. s.l.: mi-Wirtschaftsbuch, 2013.
[19] D. Schröter, P. Kuhlang, T. Finsterbusch, B. Kuhrke, and A. Verl, "Introducing Process Building Blocks for Designing Human Robot Interaction Work Systems and Calculating Accurate Cycle Times," (en), *Procedia CIRP*, vol. 44, pp. 216–221, 2016.

IOP Conf. Series: Materials Science and Engineering **235** (2017) 012006 doi:10.1088/1757-899X/235/1/012006

A Leader-path-following Formation System for AGVs with Multi-sensor Data Fusion Based Vehicle Tracking

Wen Yao[1*], Xijun Zhao[1], Yufeng Yu[2], Yongkun Fang[2], Chao Wang[1] and Tianfu Yang[1]

[1] China North Vehicle Research Institute, Beijing, China
[2] Peking University, Beijing, China

* gphbl@aliyun.com

Abstract. Caravans composed of vehicles with different functionality or trafficability raise the demand that formation system structure shall allow vehicles to deviate from the path to be followed when necessary. In this paper, a formation system is developed for autonomous ground vehicles (AGVs) who follow the path of a leader vehicle while retaining the ability of deviation from the reference path. In addition, it improves robustness of preceding vehicle localization by fusing Lidar tracking, camera tracking results with predecessor's global position within an extended Kalman filter (EKF) in case that one or more sources of preceding vehicle localization is not reliable. The system is applied on real AGV platforms and won the 3rd place in an AGV competition in China.

1. Introduction

During the past decades, formation control has received a lot of attention since it enables coordinated traversal of roads for multi-vehicle system. Techniques from adaptive cruise control (ACC), coordinated ACC (C-ACC) to advanced vehicle platooning has brought much benefits to transportation in fuel consumption, labour costs, traffic throughput and driving safety. Much of the work [1-6] in this area utilizes vehicle to vehicle (V2V) communication technique to transmit information about local control manoeuvres among caravan members so as to reduce inter-vehicle distances and decrease air drag. The results can be remarkable when vehicles in caravan share similar kinematic and dynamic characteristics because manoeuvres from preceding vehicle can be a good reference for its follower. But in many situations, a caravan may be composed of vehicles with different functionality or trafficability. In these cases, more flexible spatial constraint is required for each follower to guarantee safety and functionality. Vehicle with lower trafficability (e.g. vehicles with low chassis) may have to avoid collisions while others (e.g. all-terrain vehicle) need not.

Another group of work [7-13] on development of local control strategy is done to overcome the difficulties when only local positioning and range information to neighbouring vehicles are available. Vehicles normally maintain a formation with rigid configuration in these systems. Obstacles presenting between neighbouring vehicles are not considered.

In this paper, a formation control system is developed to achieve a caravan composed of multiple autonomous ground vehicles (AGVs) who follow the path of the leader vehicle. Follower vehicle can deviate from leader path for collision avoidance purpose. The system obtains robust preceding vehicle localization information by utilizing vehicle tracking techniques fusing Lidar, camera and GPS with an extended Kalman filter (EKF). This work is an extension based on our prior work [14] where a motion planner is implemented to drive an AGV along given road points. The AGV equipped with our formation control system presented in this paper won the 3rd place in a native competition on autonomous logistics convoy (figure 1) [15].

Figure 1. Our AGV (left) is following the leader vehicle (right) during a native AGV competition.

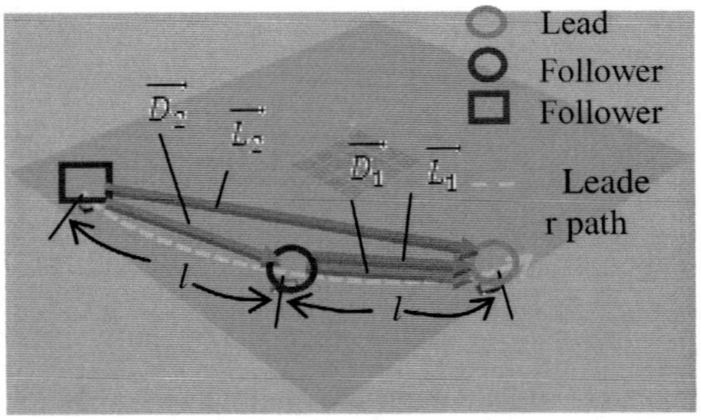

Figure 2. The definition of a leader-path following formation

Similar formation control implementation can be found in [16] where a road-following approach coupled to a tightly constrained route to be followed is presented. It enables deliberative deviation from the route for collision avoidance. The work is validated in simulation. The acquisition of accurate preceding vehicle state is not issued which is challenging in natural environment.

There are two novel contributions implemented in this work: first, a leader-path following formation system is designed to allow follower vehicles with different kinematics or trafficability to follow the leader vehicle's path with necessary lateral and longitudinal motions; second, a vehicle tracking module fusing GPS, Lidar and mono camera data with extended Kalman filter is developed for each follower to localize the preceding vehicle robustly.

The rest of the paper is organized as follows: in section 2, the formation control framework is overviewed first, followed with the generation of path-to-be-followed for each caravan member. The preceding vehicle tracking method is described in section 3 which gives a robust estimation of leader pose and distance to be kept even in severe situation where one or two sensor sources are not reliable. The simulation and experimental results on real platforms are introduced in section 4.

2. Leader-path-following formation control

In a leader-path-following formation, all the followers tries to follow the path that leader vehicle has passed. Comparing with controlling ego vehicle respect to its predecessor's current pose or passed pose sequence, leader-path following guarantees all the vehicles follow the same reference route, so that control error from one vehicle does not propagate to its successor [17].

2.1. Problem formulation

The leader-path-following formation mode implemented in this paper is indicated in figure 2. The leader vehicle is circled in red with its history path LP^* indicated with green dashed curve. Each follower i estimates the leader's state at time t as $\boldsymbol{L}_i(t)$ in its local coordinate, which is defined as:

$$\boldsymbol{L}_i(t) = \left(\Delta x_i(t), \Delta y_i(t), \Delta\theta_i(t)\right) \tag{1}$$

where $(\Delta x_i, \Delta y_i, \Delta\theta_i)$ is the pose of the leader in follower i's local coordinate. Except for the first follower right after the leader, $\boldsymbol{L}_i(t)$ may not be able to be directly measured by Lidar and camera due to occlusion. In order to make full use of visual and Lidar information, $\boldsymbol{L}_i(t)$ is calculated by:

$$\boldsymbol{L}_i(t) = \boldsymbol{L}_{i-1}(t) + \boldsymbol{D}_i(t) \tag{2}$$

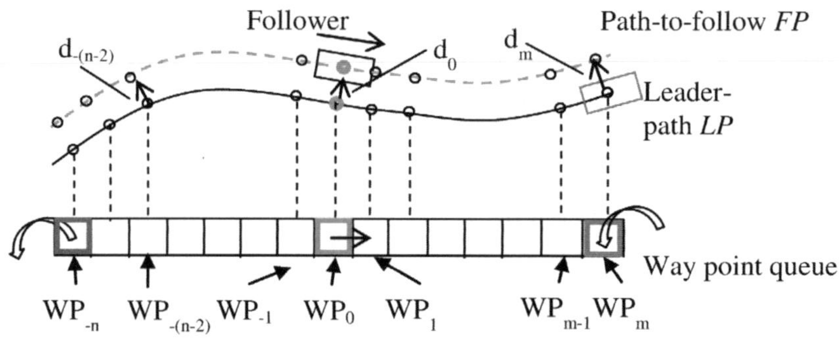

Figure 3. Way point queue and path-to-follow generation.

where $\boldsymbol{D_i(t)}$ (indicated in blue) is pose of preceding vehicle observed from ego vehicle also defined as a triplet of position and orientation difference the two neighboring vehicles. With ego pose data $\boldsymbol{P_i(t)}$ in a fixed coordinate C_i (either world coordinate or odometry coordinate) obtained from INS, each follower maintains an estimated *leader-path* LP_i

$$LP_{i,t_0 \sim t_1} = \boldsymbol{P_i}(t) + \boldsymbol{L_i}(t)|_{t_0 \sim t_1} \tag{3}$$

of leader path $LP^*_{t_0 \sim t_1}$ from leader position at t_0 to position at t_1 in C_i as its reference path to follow. The curve length along LP_i between follower i and i-1 is denoted as l_i. Note that each follower maintains a separate leader path LP_i is necessary when it is not perfectly self-localized. This is a common case in situation where global localization system is not accurate due to multi-path reflection or signal block or odometry on different followers have different precision. The error between LP^* and LP_i has to be limited in a reasonable small range for the system to run normally.

Now the formation control of each follower can be divided into two sequential sub-tasks: first, maintain an estimated leader-path LP_i and generate an FP_i as path-to-follow; second, control each follower to move along FP_i while keeping safe distance from preceding vehicle.

2.2. Path-to-follow generation

In this work, path is represented as sequence of way points. For each follower, a queue of waypoints

Algorithm 1. Maintenance of estimated leader-path LP_i and path-to-follow FP_i

Input: current leader pose WP_m, ego pose P, current gear G, lateral offset d_k
forever do
if $||WP_m - WP_{m-1}||$ > Min_step_len
  and $m <$ Max_forward_waypoint_num **then**
  LP.push_back(WP_m)
Search in LP for the respect way point of P and marked as WP_0
if Max_backward_waypoint_num == n **then**
  LP.pop_front(WP_{-n})
if G == Backward_gear **then**
  remove $\{WP_k\}_{k>0}$
    return FP = reversed $\{WP_k\}_{-n \sim 0} + \{\boldsymbol{d_k}\}_{0 \sim -n}$
else
  return $FP = \{WP_k\}_{0 \sim m} + \{\boldsymbol{d_k}\}_{0 \sim m}$

$\{WP_i\}_{-n \sim m}$ is maintained as the ego-estimated leader-path LP as depicted in figure 3 and Algorithm 1. In each iteration, a new way point WP_m (indicating the current pose of leader in red rectangle) is pushed into the end of way point queue when leader vehicle (red rectangle) is observed to have moved more than *Min_step_len* forward from last way point WP_{m-1}(line 3 to 4). When leader vehicle is too far away, point will be added. The algorithm then gets current poses P of ego vehicle and searches for the corresponding way point WP_0 in the queue (line 5). The details of this searching process is explained in our prior work [14]. To enable backward moving, this queue also stores passed way points with capacity of Max_backward_waypoint_num (line 6 to 7). Since backward moving happens when formation needs to go back to the fork in a road and choose another branch, forward queue $\{WP_k\}_{k>0}$ is emptied. The path FP for ego vehicle to follow is generated by add a sequence of lateral offset $\{\boldsymbol{d_k}\}$ to LP. The offset d can be used to control the relative position between ego and preceding vehicles. For example, let d equals to the lane width will lead the two vehicles to run in different lanes of a road. It is also possible to apply smoothing or other optimization over LP for followers of different kinematics.

2.3. Vehicle motion control along path-to-follow

Once the path FP_i to be followed by follower i is generated, steering and velocity of ego vehicle are controlled to follow FP_i and keep a safe distance from preceding vehicle. A motion planner for UGV is implemented in our prior work [15] controlling ego vehicle to move along FP_i without collision. For formation control task presented in this paper, a PID controller (see equation 4) is applied to guarantee safety with preceding vehicle. This controller tries to keep a constant distant l^* along LP (l and LP are depicted in figure 2) from predecessor who is running at a speed of v_p. The actual ego speed v_e is measured by odometry as feedback. The actual distance l from ego to its predecessor is returned by calculating the curve length along LP_i based on relative pose $\vec{D}$. The motion planner of the UGV accepts v_E as target speed and controls actuators accordingly:

$$v_E = PID\big(l - l^* + k(v_p - v_e)\big) \tag{4}$$

3. Preceding vehicle tracking

The relative pose $\vec{D}$ of preceding vehicle in ego coordinate as formulated in equation (2) in chapter 2.1 is obtained from a preceding vehicle tracking process shown in figure 4. There are 3 sources of preceding vehicle pose data from 3 different types of sensors. Global position, orientation, velocity and acceleration are obtained by integrated navigation system (INS) on preceding vehicle and transmitted to ego vehicle through communication channel. A 3D Lidar (i.e. Velodyne HDL-64E) is mounted on top of the ego vehicle and collect laser point cloud of surroundings. Vehicle detection and tracking algorithm is running on these range data and returns predecessor's state even in poor illumination. A mono camera is also set up in front of the ego vehicle and captures images ahead for a visual tracking process. Images provide more detailed texture information which helps to recognize the desired predecessor from multiple moving objects.

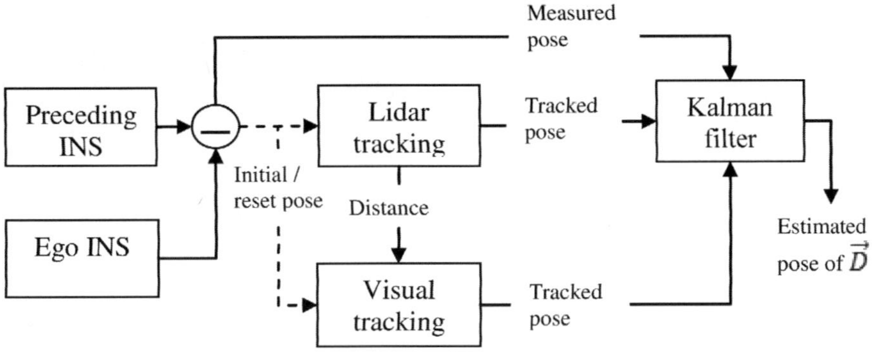

Figure 4. The work flow of preceding vehicle tracking.

3.1. Lidar tracking

The Lidar tracking process (figure 5) is initialized with preceding vehicle pose data (shown as a cyan rectangle in figure 6) received from predecessor. The region of interest (ROI) shown in red is a circle centred at the position of predecessor's GPS antenna. In the ROI, laser point height and gradient of range between laser scans are used to remove ground surface and extract foreground points, i.e. points on preceding vehicle. A hierarchical clustering process is applied to the foreground points. The clustering centre which is nearest to predicted predecessor position in last iteration is selected to be the estimation of current predecessor. The result is both passed to the fusion process and used to do pose prediction for the next tracking iteration. This tracking process runs at a frequency of 10Hz. Uniform linear motion is used as a simple motion model for pose prediction.

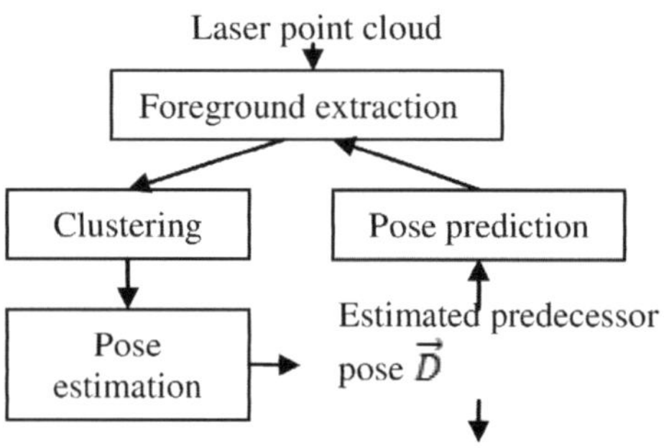

Figure 5. Vehicle tracking process based on Lidar data

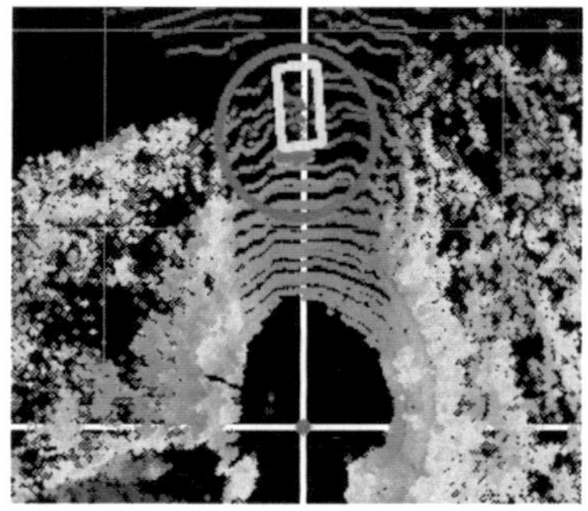

Figure 6. GPS position data (cyan) and searching window for vehicle tracking (red circle)

3.2. Visual tracking

The initialization of visual tracking also gets preceding vehicle pose from inter-vehicle communication. An initial ROI (green square in figure 7) is selected around the initial position according to the texture feature is much less sensitive to illumination conditions than color features. During on-line process, convolutions of feature template and the small windows sampled in the ROI in image feature pyramid are computed. The strongest response (white area in figure 8 right) gives the preceding vehicle position as an initial value for vehicle tracking. In each iteration during the tracking process, a vehicle detector is trained and updated by applying kernelized correlation filters (KCF) [18] algorithm, which speeds up the tracking process a lot. Since range information is not available in 2D images, the 3D Lidar used for tracking is calibrated with the camera and provides corresponding distance.

Figure 7. Visual tracking result (blue) in the initial ROI (green) and the HOG feature template used for detection (right)

IOP Conf. Series: Materials Science and Engineering **235** (2017) 012006 doi:10.1088/1757-899X/235/1/012006

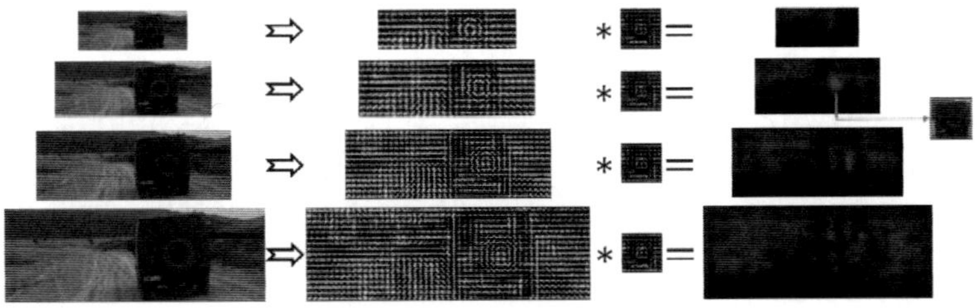

Figure 8. Searching for the strongest response (right) in the image feature pyramid (middle)

3.3. Data Fusion with Kalman filter

The 3 sources of preceding vehicle localization data are labelled with timestamps and put into the Kalman filter in chronological order. The fusion process is done in a direct EKF scheme [19]. The status of preceding vehicle in the tracking system is defined as a vector $\mathbf{X} = (x, y, \theta, v, \omega)$, with position ($x$, y), heading θ, linear and angular speed (v, ω) in odometry coordinate of the ego vehicle.

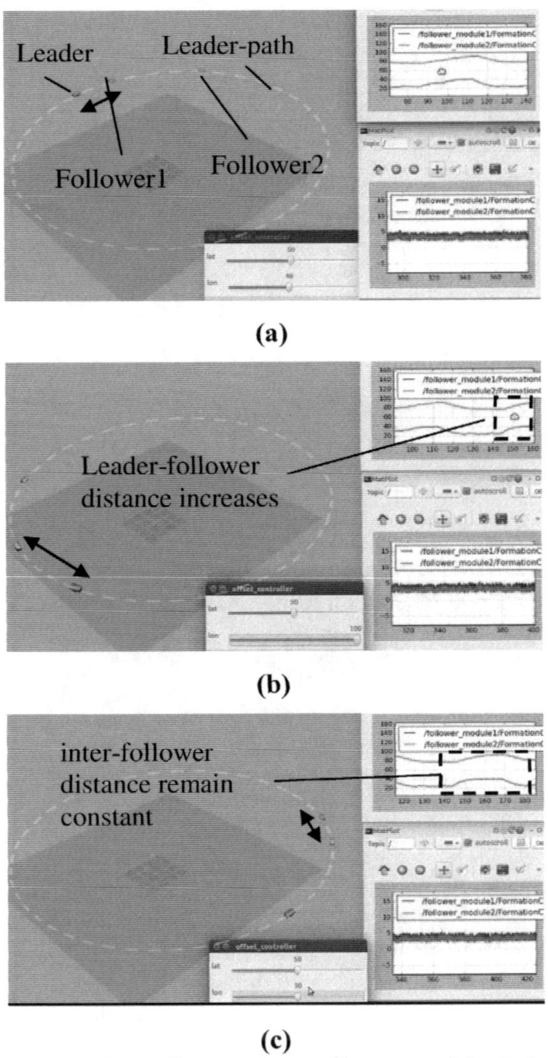

Figure 9. Followers keep distance with their predecessors while following the leader path

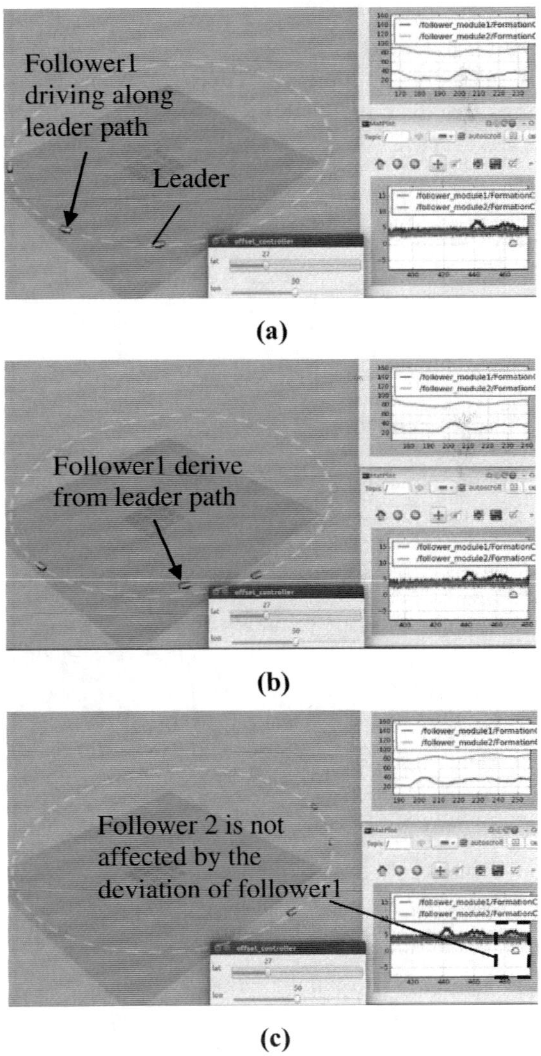

Figure 10. Deviation motion of one caravan member does not affect its successor

4. Simulation and experimental results

4.1. Simulation result

The simulation is done in an open source simulator called Gazebo [20] which provides high-performance physics engines and ROS based interfaces that can be conveniently accessed. Figure 9 and 10 show the effect on follower when preceding vehicle moves in longitudinal and lateral direction along leader path. In figure 9, the plot at top right corner of each sub figure shows the distance to the leader from two followers (follower 1 in blue, follower 2 in red). The two distances varies together keeping the difference constant. Figure 10 shows that when follower 1 deviates from the leader path (figure 10 (b)), the lateral offset of follower 2 (red plot at bottom right in each sub figure) does not change with follower 1 (blue plot). Please see the video at https://sqi.cc/c8.

4.2. Experimental results on real platform

The formation control system developed in this paper is tested on two AGV platforms. A video can be found at https://sqi.cc/c9. A one-leader-one-follower experiment is done and the path of two vehicles are drawn in figure 11 with z axis being the data frame count. The path projected to x-y plane shows that the follower path almost coincides with the leader. The time delay between two vehicles is magnified 1000 times along z-axis for a clearer view. The desired gap distance between two vehicles is set to be 30 meters. Actual gap distance (figure 12 up) and the speed (figure 12 down) of two vehicles are visualized. Figure 13 and 14 show two situation where visual tracking is not able to give the correct result. The tracking process can still provide correct localization result thanks to data fusion.

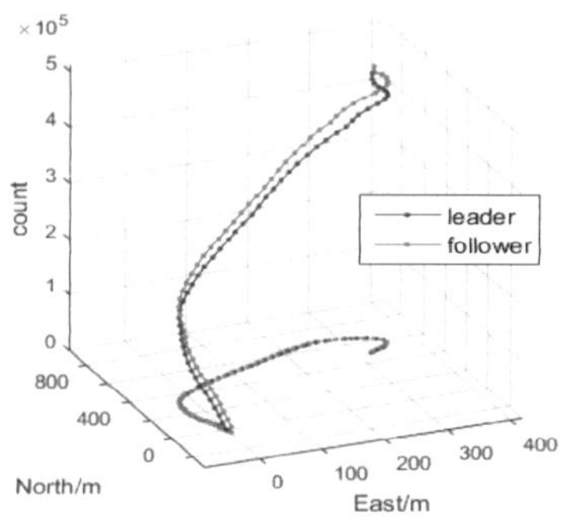

Figure 11. Leader and follower path in x-y-t space

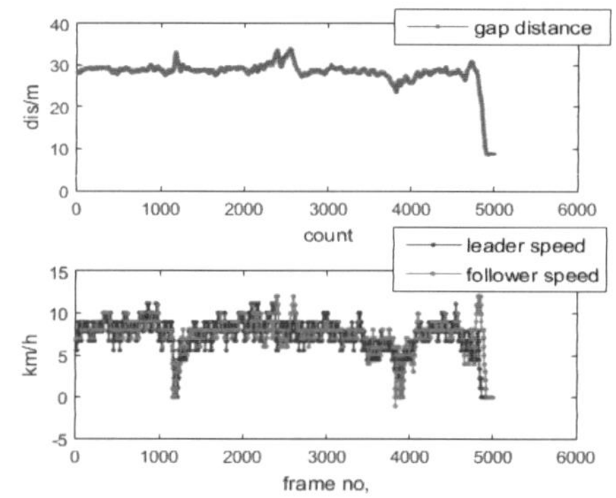

Figure 12. Longitudinal distance and speed difference between leader and follower

Figure 13. Visual tracking result (blue) is affected by the rain while GPS (cyan) and fused result (red) is correct.

Figure 14. Part of the leader vehicle exceeds image border. There is no visual tracking result. But Lidar tracking (yellow), GPS and fused result (red) are still available

IOP Conf. Series: Materials Science and Engineering **235** (2017) 012006 doi:10.1088/1757-899X/235/1/012006

5. Conclusion

A formation system is developed to achieve a caravan composed of multiple autonomous ground vehicles (AGVs) who follow the path of a leader vehicle while retaining the ability of deviation from the reference path. The system improves robustness of preceding vehicle localization by fusing Lidar, camera tracking results with predecessor's global position within an extended Kalman filter in case that one or more sources of preceding vehicle localization is not available. Future works include testing on larger scale formations and developing the ability to change formation dynamically with vehicle merging into or splitting from the formation.

Reference

[1] Tsao H S, Hall R W and Hongola B 1994 Capacity of automated highway systems: effect of platooning and barriers *PATH Research Report* UCB-ITS-PRR-93-26

[2] Cremean L and Murray R 2003 Stability analysis of interconnected nonlinear systems under matrix feedback *Proc. IEEE Int. Conf. on Decsion and Control* vol 3 pp 3078-3083

[3] Xiao L and Feng G 2011 Practical string stability of platoon of adaptive cruise control vehicles *IEEE Trans. on Intelligent Transportation System* vol 12 no 4 pp 1184-1194

[4] Karagiannis G *et al.* 2011 Vehicular networking: a survey and tutorial on requirements, architectures, challenges, standards and solutions *IEEE Commun. Surveys &Tutorials* vol 4 no 13 pp 584-616

[5] Alam A A *et al.* 2010 An experimental study on the fuel reduction potential of heavy duty vehicle platooning *Proc IEEE Int. Conf. on Intelligent Transportation Systems* pp 306-311

[6] Vinel A, Lan L and Lyamin N 2015 Vehicle-to-vehicle communication in C-ACC/platooning scenarios *IEEE Communications Magazine* vol 53 no 8 pp 192-197

[7] Mesbahi M and Egestedt M 2010 *Graph theoretic methods in multi agent networks* Princeton University Press

[8] Keviczky T Borrelli F and Balas G 2006 Decentralized receding horizon control for large scale dynamically decoupled system *Automatica* vol 42 no 12 pp 2105-2115

[9] Liu T and Jiang Z P 2013 A nonlinear small-gain approach to distributed formation control of nonholonomic mobile robots *Proc. IEEE Conf. American Control* pp 3051-3056

[10] Ji M and Egerstedt M 2007 Distributed coordination control of multiagent systems while preserving connectedness. *IEEE Trans. on Robotics* vol 23 no 4 pp 693-703

[11] Smith B S Wang J Egerstedt M and Howard A 2009 Automatic formation deployment of decentralized heterogeneous multi-robot network with limited sensing capabilities *Proc. IEEE Int. Conf. on Robotics and Automation* pp 2130-2135

[12] Zhang F and Haq S 2008 Boundary following by robot formations without gps *Proc. IEEE Conf. on Robotics and Automation* pp 153-157

[13] Gowal S Falconi R and Martinoli A 2010 Local graph-based distributed control for safe highway platooning *Proc. IEEE Int. Conf. on Intelligent Robots and Systems* pp 6070-6076

[14] Zhao X Liu J Zhu S Zhu L and Wang H 2016 A motion planning method for autonomous vehicles *Proc. Int. Conf. on Control and Automation*

[15] Chen L 2016 Unmanned ground systems to multitask for Chinese army *China Daily* http://www.chinadaily.com.cn/china/2016-10/18/content_27098480.htm

[16] Ono M Droge G Grip H Toupet O Scrapper C and Rahmani A 2015 Road-following formation control of autonomous ground vehicles *Proc. IEEE Conf. on Decision and Control* pp 4714-4721

[17] Barooah P and Hespanha J P 2005 Error amplification and disturbance propagation in vehicle strings with decentralized linear control *Proc. IEEE Conf. on Decision and Control* pp 4964-4969

[18] Henriques J F Caseiro R Martins P 2015 High-speed tracking with kernelized correlation filters *IEEE Trans. on Pattern Analysis & Machine Intelligence* vol 37 no 3 pp 583-596

[19] Sasiadek J Z and Hartana P 2000 Sensor data fusion using Kalman filter *Proc. Int. Conf. on Information Fusion* vol 2 pp WED5/19-WED5/25

[20] Gazebo simulator http://www.gazebosim.org

IOP Conf. Series: Materials Science and Engineering **235** (2017) 012007 doi:10.1088/1757-899X/235/1/012007

Generating Poetry Title Based on Semantic Relevance with Convolutional Neural Network

Z Li[1], K Niu[1] and Z Q He[1]

[1]Key Laboratory of Universal Wireless Communications, Ministry of Education, Beijing University of Posts and Telecommunications, No.10 Xitucheng Road, Haidian District, Beijing, China

jaylee1992@foxmail.com

Abstract. Several approaches have been proposed to automatically generate Chinese classical poetry (CCP) in the past few years, but automatically generating the title of CCP is still a difficult problem. The difficulties are mainly reflected in two aspects. First, the words used in CCP are very different from modern Chinese words and there are no valid word segmentation tools. Second, the semantic relevance of characters in CCP not only exists in one sentence but also exists between the same positions of adjacent sentences, which is hard to grasp by the traditional text summarization models. In this paper, we propose an encoder-decoder model for generating the title of CCP. Our model encoder is a convolutional neural network (CNN) with two kinds of filters. To capture the commonly used words in one sentence, one kind of filters covers two characters horizontally at each step. The other covers two characters vertically at each step and can grasp the semantic relevance of characters between adjacent sentences. Experimental results show that our model is better than several other related models and can capture the semantic relevance of CCP more accurately.

1. Introduction

CCP is a great treasure of Chinese culture. According to the structural, rhythmical and tonal patterns, CCP can be divided into many genres. Among all the genres, the quatrain perhaps is the most popular one. There are four lines in the quatrain and each line has five or seven characters. An example of a 5-char quatrain written by Li Bai, a famous poet in Tang Dynasty of China, is shown in figure 1.

《静夜思》
《In the Still of the Night》
窗前明月光，
I descry bright moonlight in front of bed.
疑是地上霜。
I suspect it to be hoary frost on the floor.
举头望明月，
I watch the bright moon, as I tilt back my head.
低头思故乡。
I yearn, while stooping, for my homeland more.

Figure 1. An example of a 5-char quatrain.

In recent years, there are several studies using deep learning to write CCP, for examples, the RNNPG model[1] with three neural networks and the encoder-decoder model[2] with a recurrent neural network (RNN) as the encoder and another RNN as the decoder. Given some keywords, those models can generate poems line by line meeting the structural, rhythmical, tonal constraints. However, it is still difficult to automatically generate the title of CCP. The difficulties mainly come from the differences between the CCP and the modern Chinese articles, which will be described in detail in section 2.

In this paper, we focus on making machines grasp the semantic meaning of the CCP and automatically generate a semantic relevance title. We treat this as an automatic text summarization problem. The research of automatic text summarization is very active in recent years and the encoder-decoder model is widely used recently. Rush, Chopra and Weston[3] first introduce the encoder-decoder model into the automatic text summarization problem and they name their model as ABS. [4] uses a RNN for encoding and attention mechanism is used in the RNN encoder. Many techniques have been proposed based on the encoder-decoder model to improve the summarization result, such as feature-rich encoder[5], minimum risk training[6] and copy mechanism[7]. Our work is based on the encoder-decoder model and we improve the structure of the encoder, which can capture the semantic relevance in poems and is more suitable for the title generation of CCP.

The paper is organized as follows: In section 2, we describe the problem of poem title generation. The details of poem title generation (PTG) model we proposed are introduced in section 3. At last, the evaluation result and conclusion are shown in section 4 and 5 respectively.

2. Problem description

In this paper, we focus on the task of automatically generating the title of CCP. In this task, each input is a quatrain, which is a famous genre of CCP. We first train word2vec[8] model using our dataset to represent each characters as a k-dimensional vector. Let $\boldsymbol{x}_{ij} \in \mathbb{R}^k$ be the k-dimensional column vector of the j-th character in the i-th sentence. Let m be the total number of sentences in one quatrain and n be the total number of characters in one sentence, then a quatrain can be represented as a matrix $\boldsymbol{X} \in \mathbb{R}^{m(kn)}$ as below:

$$\boldsymbol{X} = \begin{bmatrix} \boldsymbol{x}_{11}^T & \boldsymbol{x}_{12}^T & \cdots & \boldsymbol{x}_{1n}^T \\ \boldsymbol{x}_{21}^T & \boldsymbol{x}_{22}^T & \cdots & \boldsymbol{x}_{2n}^T \\ \vdots & \vdots & \ddots & \vdots \\ \boldsymbol{x}_{m1}^T & \boldsymbol{x}_{m2}^T & \cdots & \boldsymbol{x}_{mn}^T \end{bmatrix} \tag{2.1}$$

Given the content of a poem $\boldsymbol{X}$, the model needs to generate the title $\boldsymbol{y} = [y_1, y_2, ..., y_k, ..., y_K]$ with maximum probability, where y_k is the k-th character in the title. The process can be denoted as below:

$$\boldsymbol{y} = arg\max_y p(\boldsymbol{y} \mid \boldsymbol{X}) \tag{2.2}$$

This task can be taken as an automatic text summarization problem. However, there are several differences between our problem and the summarization problem of modern articles. First, the words in CCP are very different from the words in the modern Chinese and there are no valid word segmentation tools we can use. Just using characters is not enough and we need to make use of the information of words. Second, there are structural, rhythmical, tonal constraints in CCP, so it's better for our model to take advantage of that information. Third, the title of CCP is extremely short (usually less than five characters), so the popular evaluation metrics such as ROUGE[9] are not suitable for this problem. All the differences bring new challenges to the poem title generation problem. For the first and second difficulties, we proposed our poem title generation (PTG) model. We also introduce a new kind of evaluation metric which can reflect the semantic differences between titles to solve the third difficulty.

3. The PTG model

In this section, we first describe the semantic features of CCP, and then the framework for poem title generation is presented. After that, we show the implementation details of our model and how the model can capture the semantic features.

3.1. The semantic features of CCP

The coherence is an important feature of CCP. When writing a CCP, the poet pays attention to not only the semantic relevance of characters in one sentence, but also the semantic relevance of characters between adjacent sentences. So the semantic features of CCP are mainly reflected in two aspects, the intra-sentence semantic relevance and the inter-sentence semantic relevance.

The intra-sentence semantic relevance of characters means some characters usually appears together in one sentence and form words. In CCP, the words usually consist of two characters. In figure 1, "moon" and "light" in the first sentence of the quatrain are an example of the intra-sentence semantic relevance, which form the word "moonlight" together.

The inter-sentence semantic relevance of characters is the antithesis used between two adjacent sentences in the poem. If two characters appear in the same position of two adjacent sentences and the meanings of the two characters are related to each other, then there is inter-sentence semantic relevance between the two characters. In the quatrain in figure 1, the first character "raise" of the third sentence and the first character "stoop" of the fourth sentence are an example of the inter-sentence semantic relevance.

There are difficulties for a model to capture the two kinds of semantic features. The words used in the CCP have many differences with the model Chinese words and there are no valid segmentation tools. So our model should learn the information of words directly from the characters. In addition, the inter-sentence semantic relevance is reflected from the characters in the same position of adjacent sentences, so the input form of our model should not simply be a sequence of characters, which usually is the input form of traditional encoder-decoder models. We will show how our model overcomes those difficulties and captures both the two kinds of semantic features of CCP in the below sections.

3.2. The framework for poem title generation

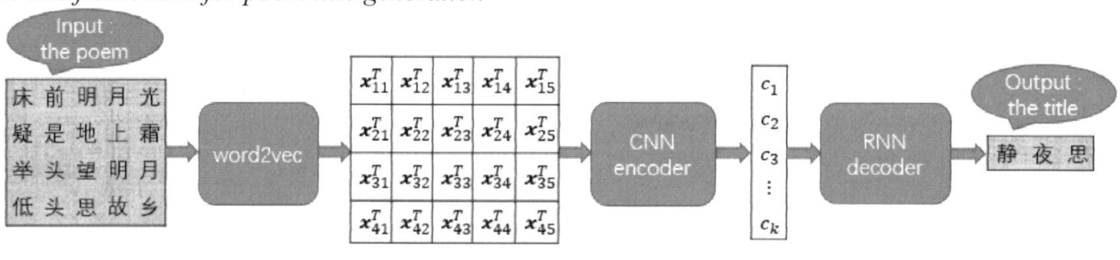

Figure 2. The framework for poem title generation

The poem title generation problem is similar to the automatic text summarization problem. Given an input, which is the quatrain in our problem, the PTG model needs to generate an output which represents the main ideas of the input and is the title of the quatrain here. The entire framework for poem title generation is shown in Figure 2. The title of a CCP can be generated through three steps in our PTG model. Step 1, by looking up the word embedding matrix trained by word2vec model, each character in the poem is transformed into a k-dimensional vector. Step 2, the model encoder which is a CNN encodes the poem into a fixed length vector, which represents the semantic meaning of the original poem. Step 3, the model decoder which is a RNN generates the title character by character as the above vector being the context.

Our model is similar to the encoder-decoder model raised in [10], but the encoder is tailored CNN for our task. We will describe the details of the CNN encoder in the next section.

3.3. The CNN encoder

The encoder is to encode the input into a fixed length vector. Here, each input is a quatrain $\boldsymbol{X}$ shown in equation (2-1). The quatrain has four sentences and each sentence consists of five or seven characters, so here m is 4 and n is 5 or 7. Because of the special structure of quatrains, we can take quatrains as squared pictures.

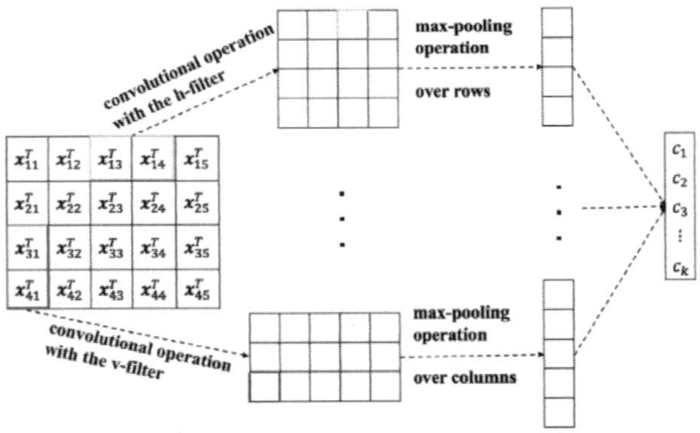

Figure 3. The operations in CNN encoder

Our encoder is a convolutional neural network. The convolutional operation is associated with the filter $\boldsymbol{W} \in \mathbb{R}^{ab}$. To capture the two kinds of semantic features mentioned above, we design two kinds of filters with different shapes. The first kind of filter, which is drawn yellow in figure 3, is the horizontal filter (h-filter) denoted as $\boldsymbol{W}_h$ with a=1 and b=2k. Another kind of filter, which is drawn red in figure 3, is the vertical filter (v-filter) denoted as $\boldsymbol{W}_v$ with a =2 and b = k. Then the convolutional operations are induced as below to get an intra-sentence features c_{ij}^{h} and an inter-sentence feature c_{ij}^{v}:

$$c_{ij}^{h} = f(\sum_{p=1}^{a}\sum_{q=1}^{b}(\boldsymbol{W}_h \odot \boldsymbol{X}_{i,j:j+1})_{pq} + b_h) \tag{3.1}$$

$$c_{ij}^{v} = f(\sum_{p=1}^{a}\sum_{q=1}^{b}(\boldsymbol{W}_v \odot \boldsymbol{X}_{i:i+1,j})_{pq} + b_v) \tag{3.2}$$

where, $\boldsymbol{X}_{i,j:j+1} = \begin{bmatrix} \boldsymbol{x}_{i,j}^{T} & \boldsymbol{x}_{i,j+1}^{T} \end{bmatrix}$ and $\boldsymbol{X}_{i:i+1,j} = \begin{bmatrix} \boldsymbol{x}_{i,j}^{T} \\ \boldsymbol{x}_{i+1,j}^{T} \end{bmatrix}$ are parts of the input $\boldsymbol{X}$. The mark $\odot$ indicates the element-wise multiplication. $b_h \in \mathbb{R}$ and $b_v \in \mathbb{R}$ are the bias term. f is a non-linear function such as the hyperbolic tangent.

The h-filter covers two adjoining characters in one sentence. After training on the dataset, the h-filter can learn the commonly used words of two characters without word segmentation, so the intra-sentence semantic relevance is captured by our model. The v-filter covers two characters in the same position of two adjacent sentences, so it can capture the inter-sentence semantic relevance. When applying the h-filter to row i of $\boldsymbol{X}$, we can get an intra-sentence feature list denoted by $\boldsymbol{c}_i^h = \begin{bmatrix} c_{i,1}^{h} & c_{i,2}^{h} & \cdots & c_{i,n-1}^{h} \end{bmatrix}$.

After the convolutional operation, we use the max-pooling operation. The max-pooling operation is taking the max value of each feature list which is denoted as $c_i^h = max(\boldsymbol{c}_i^h)$. The aim of max-pooling operation is to capture the most important feature of each feature list. So after the convolutional and max-pooling operations, a specific h-filter can generate an intra-sentence feature list

$\boldsymbol{c}^h = \begin{bmatrix} c_1^h & c_2^h & \dots & c_m^h \end{bmatrix}$. Similarly, a specific v-filter can generate an inter-sentence feature list $\boldsymbol{c}^v = \begin{bmatrix} c_1^v & c_2^v & \dots & c_n^v \end{bmatrix}$.

We have described the process of how each filter can generate each list of features. Our CNN model uses multiple h-filters and v-filters to obtain multiple feature lists. We concatenate all feature lists to form one k-dimensional vector $\boldsymbol{c} \in \mathbb{R}^k$, which is the representation of the whole quatrain. Our model can capture both the two kinds of semantic features mentioned above, so the vector $\boldsymbol{c}$ can represent the semantic meaning of the corresponding quatrain.

4. Experiments and results

The training dataset we use is 40047 quatrains with titles. We compare our model with two baseline systems, which are both encoder-decoder models. The first is RNN encoder-decoder model raised in [11]. The second one[4] is similar with the first one, but uses the attention mechanism[12] in the encoder.

4.1. Evaluation metrics

The traditional evaluation metrics for automatic text summarization task such as ROUGE can only reflect the surface similarity between the standard title and the generated title. This is not suited for our poem title generation task. As we mentioned before, the title of CCP usually has less than five characters and directly comparing the surface similarity is ineffective. The semantically similar titles may have low scores under the traditional evaluation metric. In order to reflect the semantic similarity between the standard title and the generated title, we proposed two kinds of evaluation metrics. Let $\boldsymbol{G} = \{(\boldsymbol{X}^{(1)}, \boldsymbol{y}^{(1)}, \boldsymbol{y}'^{(1)}), (\boldsymbol{X}^{(2)}, \boldsymbol{y}^{(2)}, \boldsymbol{y}'^{(2)}), \cdots, (\boldsymbol{X}^{(M)}, \boldsymbol{y}^{(M)}, \boldsymbol{y}'^{(M)})\}$ denotes the standard poem-title pairs in the test dataset and the corresponding generated titles. Here $\boldsymbol{y} = \{\boldsymbol{y}_1, \boldsymbol{y}_2, \cdots, \boldsymbol{y}_j, \cdots, \boldsymbol{y}_K\}$ denotes the standard title of the poem and $\boldsymbol{y}' = \{\boldsymbol{y}'_1, \boldsymbol{y}'_2, \cdots, \boldsymbol{y}'_j, \cdots, \boldsymbol{y}'_{K'}\}$ denotes the model generated title. $\boldsymbol{y}_j \in \mathbb{R}^k$ and $\boldsymbol{y}'_j \in \mathbb{R}^k$ are vectors, which are trained by word2vec model, of j-th characters in the titles. The vectors trained by the word2vec model can represent the semantic meaning of characters and the distance such as cosine distance between vectors can show the semantic similarity between corresponding characters. The first evaluation metric s_1 we proposed is calculated as below:

$$s_1(\boldsymbol{y}_T, \boldsymbol{y}'_T) = s(\boldsymbol{y}_T, \boldsymbol{y}'_T) = \frac{\boldsymbol{y}_T \cdot \boldsymbol{y}'_T}{\|\boldsymbol{y}_T\| \ \|\boldsymbol{y}'_T\|} \tag{4.1}$$

where the vectors of all characters in the title are averaged to formulate the standard title vector $\boldsymbol{y}_T \in \mathbb{R}^k$ and the generated title vector $\boldsymbol{y}'_T \in \mathbb{R}^k$. $s(\cdot)$ is the cosine similarity function between two vectors.

The s_1 can reflect the semantic differences between different titles. However, with the fact that the title of CCP is usually very short, directly comparing the title is not very effective, so we propose the second evaluation metric. Instead of comparing the standard titles and generated titles directly, the second evaluation metric compares the titles with the corresponding poems. The title represents the central thought of the poem, so a good title is usually closely related to the poem. Inspired by those ideas, we propose the s_2 calculated as below:

$$s_2(\boldsymbol{y}_T, \boldsymbol{y}'_T, \boldsymbol{x}_P) = \frac{s(\boldsymbol{y}'_T, \boldsymbol{x}_P)}{s(\boldsymbol{y}_T, \boldsymbol{x}_P)} \tag{4.2}$$

where the vectors of all characters in the poem are averaged to formulate the poem vector $\boldsymbol{x}_P \in \mathbb{R}^k$.

4.2. Evaluation results

Our test dataset is 1300 quatrains with titles. After the titles are generated by models, we calculate the s_1 score and the s_2 score of each generated titles. First, we compare the mean and the median of both s_1 and s_2 between different models. The results are shown in table 1.

Table 1. Evaluation results.

Model	Mean of s_1 (%)	Median of s_1 (%)	Mean of s_2 (%)	Median of s_2 (%)
RNN encoder-decoder	37.7	35.1	89.4	85.1
RNN encoder-decoder + attention	38.2	35.0	91.1	86.4
PTG (our model)	40.4	37.0	95.4	91.4

From table 1, we can observe that the PTG we proposed is better than the other two models under both s_1 and s_2. This indicates the CNN decoder we design is effective and the convolutional neural network is more suitable for the structure of quatrains. It can not only extract features in one line but also can learn features between lines.

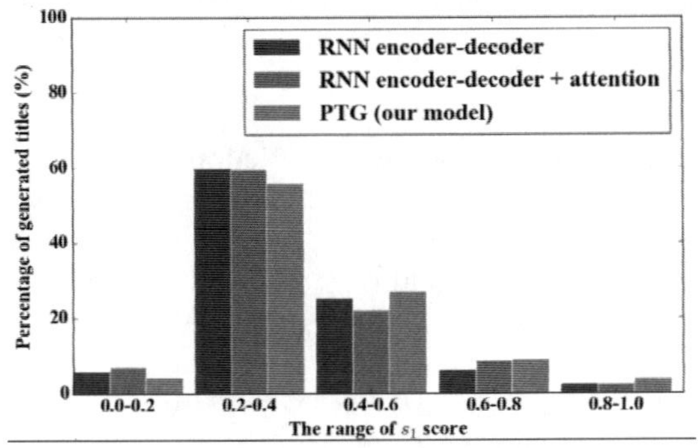

Figure 4. The percentage of generated titles in different s_1 score ranges

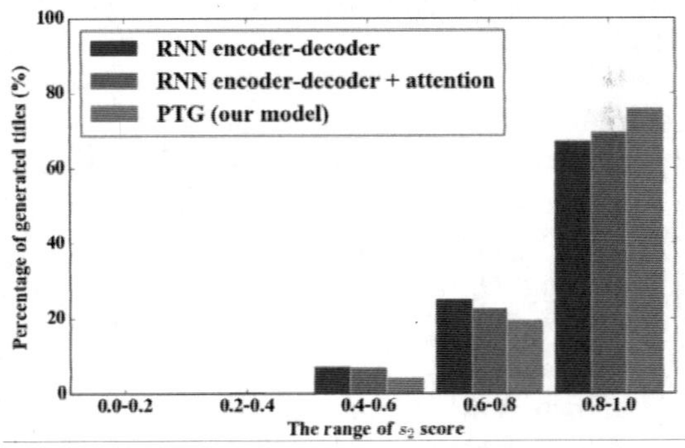

Figure 5. The percentage of generated titles in different s_2 score ranges

Second, we divided the s_1 score and the s_2 score into five ranges (0.0-0.2, 0.2-0.4, 0.4-0.6, 0.6-0.8, 0.8-1.0) and analyse the percentage of generated titles of different models in each range. The results are shown in figure 4 and figure 5 respectively. From figure 4 and figure 5, we can observe that the percentage of titles generated in PTG model is lower in low score ranges (0.0-0.4 in figure 4 and 0.0-0.8 in figure 5) and higher in high score ranges (0.6-1.0 in figure 4 and 0.8-1.0 in figure 5). This indicates that the proposed PTG model is better in generating semantic relevance titles.

4.3. Case study

To demonstrate the effectiveness of our model more specifically, we analyze the different titles generated by different models corresponding to the same quatrain. The quatrain with the standard title is shown in Figure 6. It is a poem that praises the beauty of willow.

《咏柳》
《The willow》
碧玉妆成一树高，
Emerald fashioned into a tall tree,
万条垂下绿丝绦。
Hung with myriad silk braids of green.
不知细叶谁裁出，
Who tailored the dainty leaves?
二月春风似剪刀。
The snipping wind of early spring.

Figure 6. A quatrain with standard title.

Table 2. The titles generated by different models

Model	Title
RNN encoder-decoder	The spring
RNN encoder-decoder + attention	The early spring
PTG (our model)	The willow

The titles generated by different models is shown in table 2. From table 2, we can observe that the title the RNN encoder-decoder model generated is "The spring" and the title the RNN encoder-decoder model with attention mechanism generated is "The early spring". They both focus on the topic "spring". Although the season described in the poem is the spring, but that is not the key idea this poem focuses on. In contrast, the PTG model titled this poem "The willow". Our model gets the main idea of this poem although there is no word "willow" in the content of the poem.

5. Conclusions

In this paper, we propose an encoder-decoder model for automatically generating the title of CCP. The encoder is a CNN and it is to encode the poem into a fixed length vector. The decoder is a RNN that generates the title character by character as the above vector being the input. We also propose an evaluation metric that is more suitable for evaluating our task. Experimental results show that our model is outperformed than the other related models. The future work of this paper may be trying to combine the attention mechanism with our CNN encoder to make further improvement.

Acknowledgments

This work is supported by the National Natural Science Foundation of China (61171099 and 61671080).

References

[1] Zhang X and Lapata M, "Chinese Poetry Generation with Recurrent Neural Networks," *Proceedings of the 2014 Conference on Empirical Methods in Natural Language Processing*, 2014, pp. 670–680.

[2] Wang Q, Luo T and Wang D, "Can Machine Generate Traditional Chinese Poetry? A Feigenbaum Test," *Advances in Brain Inspired Cognitive Systems: 8th International Conference*, 2016.

[3] Rush A M, Chopra S and Weston J, "A Neural Attention Model for Abstractive Sentence Summarization," *Proceedings of the 2015 Conference on Empirical Methods in Natural Language Processing*, 2015, pp. 379–389.

[4] Lopyrev K, "Generating news headlines with recurrent neural networks," *Computing Research Repository*, 2015.

[5] Nallapati R, Zhou B, Nogueira C, Gulcehre C and Xiang B, "Abstractive Text Summarization using Sequence-to-sequence RNNs and Beyond," *In Proceedings of the 20th SIGNLL Conference on Computational Natural Language Learning*, 2016, pp. 280–290.

[6] Ayana, Shen S, Zhao Y, Liu Z and Sun M, "Neural headline generation with minimum risk training," *arXiv preprint arXiv:1604.01904*, 2016.

[7] Zeng W, Luo W, Fidler S and Urtasun R, "Efficient Summarization with Read-Again and Copy Mechanism," *arXiv preprint arXiv:1611.03382*, 2016.

[8] Mikolov T, Chen K, Corrado G and Dean J. "Efficient estimation of word representations in vector space," *International Conference on Learning Representations Workshop*, 2013.

[9] Lin C Y, "Rouge: A package for automatic evaluation of summaries," *Text summarization branches out: Proceedings of the ACL-04 workshop*, 2014, Vol. 8.

[10] Sutskever I, Vinyals O and Le Q V, "Sequence to sequence learning with neural networks," *Advances in neural information processing systems*, 2014.

[11] Cho K, Merrienboer B, Gulcehre C, Bahdanau D, Bougares F, Schwenk H and Bengio Y, "Learning Phrase Representations using RNN Encoder–Decoder for Statistical Machine Translation," *Proceedings of the 2014 Conference on Empirical Methods in Natural Language Processing*, 2014, pp. 1724–1734.

[12] Bahdanau D, Cho K and Bengio Y, "Neural machine translation by jointly learning to align and translate," *International Conference on Learning Representations*, 2015.

IOP Conf. Series: Materials Science and Engineering **235** (2017) 012008 doi:10.1088/1757-899X/235/1/012008

Professional and Social Media Sites (SMSs): Motives and Positive Values of Accommodating Social Media Sites (SMSs) in Teaching Practices according to Indonesian Professional Educators: A Case Study in Two Indonesian Higher Educational Institutions

J Y Luke[1, a] and Y L Billy[2, b]

[1]Language Center, English Department, Faculty of Humanities, Bina Nusantara University, Palmerah – Jakarta Barat 11380, Indonesia.
[2]General English Course Department, Faculty of Communication, Multimedia Nusantara University, Gading Serpong, Indonesia,

[a]jluke@binus.edu; [b] johnbilly@umn.ac.id

Abstract. In millennium era, the proliferating Social Media Sites (SMSs) has not only brought increasing demands for all humans, but also creates positive values, specifically for the professional educators or lecturers in any ages. This study envisages the positive values of accommodating Social Media Sites (SMSs) in teaching practices according to the professional educators. Thirty professional educators, i.e. the lecturers, from two universities (i.e. Multimedia Nusantara University and Bina Nusantara University) has participated in this study. The data was collected from the survey by means of questionnaires, analysed using percentages, and exposed the results descriptively. The findings reflected that the positive values of accommodating Social Media Sites in teaching practices were to develop social skills and improve academic skills. However among the two values, the latter was highly influencing the professional educators because of the four reasons: enabling to do tutorial lessons, providing online discussion space with experts or guest lecturers, assisting in doing peer-review and peer-editing, and enhancing the receptive skills, the productive skills, and also the critical thinking skills of the users in SMSs, especially the professional educators or lecturers. Thus, accommodating Social Media Sites (SMSs) in teaching practices is essential for professional educators in Indonesia.

1. Introduction

The advance technology of ICT has influenced the higher education system dramatically. Demands to find information and build relations in a quick access are getting increased in the globalized society. Moreover, the challenges to provide skilful and competent graduates equipped with technology knowledge has overwhelmed the educational stakeholders in the last two decade. As results, the competitions for serving high quality standard of educational system supported by the advanced ICT technology becomes the main priority for the higher educational institutions, particularly in Indonesia.

As the growth of ICT role in educational sector, most of university in Indonesia turn to develop flexible and dynamic education system by integrating Social Media Sites (SMss). The use of Facebook, Twitter, LinkedIn, What's Up, and YouTube are often found used in every classroom as the main teaching aids or even as supplementary tools for the learning engagement. In this process, it is confirmed that there is a shifting paradigm in teaching practices, from the traditional approach to modern approach. Manca and Ranieri [1] point that "social network sites are progressively gaining attention in relation to school and higher education practices and pedagogies with important implications for changing and adjusting teaching and learning". In this case, the excessive application of online networking sites predominantly improve the teaching pedagogies that constitutes to better outcomes of teaching practices.

Meanwhile, the integration of Social Media Sites (SMSs) do not always postulate academically profits for the users especially the professional educators, i.e. the lecturers. This is due to the fact that the nature of online networking sites is everything gets shared. Thus, in some cases the higher institution prevents the ONSs users from accessing the less important sites to their academic benefits. However, the policy abides the needs of retaining information to foster the learning. The educators will have less chance to support what the students' wants to learn and how they will find answers in their learning process. Srivastava [2] supports Greenhow's points that 'social networking sites offer more than just social fulfilment or professional networking, also have implication for educators, who now have a vast opportunity support what students are learning on the web sites". In other words, the professional educators, i.e. the lecturers, are really suggested to consider the potentials of expanding online networking sites to teaching practices since it allows developments on academic skills and socials skills.

Based on the facts mentioned above, it is necessary for identifying positive values of Social Media Sites applied in teaching practices in spite of the rigorous scams to uncover the potentials advantages for the professional educators as well as motives underlying the applications of the SMSs. Thus, this study frames its objectives into three considerations as follows:

1. To find out types of Social media Sites regularly used by the Indonesian professional educators in teaching practices at the two higher educational institutions.
2. To verify the Indonesian professional educators' motives of incorporating Social Media Sites in the teaching practices at two higher educational institutions.
3. To investigate the positive values after incorporating Social Media Sites in teaching practices at higher educational institutions in Indonesia.

2. Theories underpinning

The present of artificial intelligence products in the educational field have played important role in today's higher education system. Since it emerged in every subfield of higher institutions, including in teaching and learning field, it has already diffused the knowledge and information that support the development of educational quality and the capacity of people works in educational environment as the professional educators or lecturers. As a result, the advance of ICT has yielded a shifting paradigm of teaching practices in higher education contexts. Traditional way of teaching has moved to mobile compatible learning through social media platforms as Facebook, Twitter, Blogs, You Tube, etc. Even the platform of Learning Management System (LMS) provided by higher institutions has been left behind by most of academics in the university.

Nowadays, with the expansion of machine learning in every aspect of human life, majority of the professionals in educational environment seem disengaged or uninterested to use the previous artificial intelligence platform as LMS (learning Management System) more frequent beyond the campus environment. The reasons for that come from the fact of the limited access of network of LMS, lack of sociability functions, and most importantly the LMS could not support the necessity of browsing research and collaboration information that become a part of main jobs of the academics and staffs of the universities. Sanchez-Franco [3] mentions that "using existing commercial LMSs, however, often has practical constraints". In short, at present LMS are also creating substantial problems for the users.

Thus, by the progression of technology in information and communication field as Web 2.0, the applications of social media have aroused recent interests within the education community, particularly the professional educators. They become more active users of some popular social media sites, such as Facebook, Twitter, and Instagram. As Boyd and Ellison says that "Facebook, Twitter, LinkedIn, etc. are part of the Web 2.0 social media, characterized by the creation and sharing of the user-generated materials with the common interest" [4]. So, it is common that Web 2.0 platforms as Facebook is typically designed for social networking among particular users that have similar common interest.

In Indonesia, people view that internet users are also social media users. Currently, Indonesia is the fastest growing country in relation to the number of internet users today compared to 2016. Nguyen [5] says, "Indonesia is almost 3 times over the global average with a 51% growth year-on-year". In addition based on Nguyen's finding the country (Indonesia) also comes in 4th place when it comes to the country with the largest number of active users of Facebook, only behind Brazil, India and the United States" [5]. Therefore, social media as Facebook is so common and popular among all youth Indonesian people, including novice educators in higher institutions like the universities or colleges. As a results, a lot of educators in Indonesia extend the application of social media sites in the teaching practices.

However, the idea of accommodating social media in teaching practices is not always successfully supported by the institutions and the educators, exclusively for those who live in rural areas. These professionals are reluctant to use any social media. The main reason for not using any social media because of the educators' resistance to change their habit and cultural values. In Indonesia, for example, face-to-face interaction is highly valued in academic context than a distant communication because it enables two-way communication process between students and the teacher physically. Hence, online learning is sometimes viewed sceptical because it enforces the professional educators to own personal computer, provide network access, and also technical skills to operate the digitalized system. In sum, promoting social media in teaching practices is too overpriced.

Furthermore, the free limitation of time and space when the professionals work with the social media sites may lead to inefficient behaviours. Hence, the professionals in tertiary institutions are welcome to new advance technology, but when it comes to the implementation, they agree that skills and support utilities are the vital constraints. As a result, the implementation of the robust technology of information and communication for higher education institution in Indonesia are still debatable in terms of motives and positive values of accommodating social media sites in teaching practices.

3. Methods
This study is a descriptive analysis study that focus on motives and values for accommodating Social Media Sites in teaching practices by professional educators at the two universities in Indonesia. The detailed research methods are as follows:

3.1. Subjects
The study involved 30 participants from two university academicians: fifteen lecturers from Multimedia Nusantara Universities and fifteen lecturers from Bina Nusantara Universities. All of t respondents were randomly selected, excluding the faculties and length of working factors.

3.2. Instruments
The instrument used was questionnaires. The questionnaire consists of two primary sections: 1) types of Social Media frequently used and views on the positive values gained after using the Social Media Sites in teaching practices; 2) reasons that influence positive values of applying Social Media in teaching practices. The data were analysed using percentage and descriptive analysis.

4. Results
The results of the study are described as follows:

4.1. Types of Social Media Commonly Incorporate in Teaching Practices at Higher Institutions in Indonesia

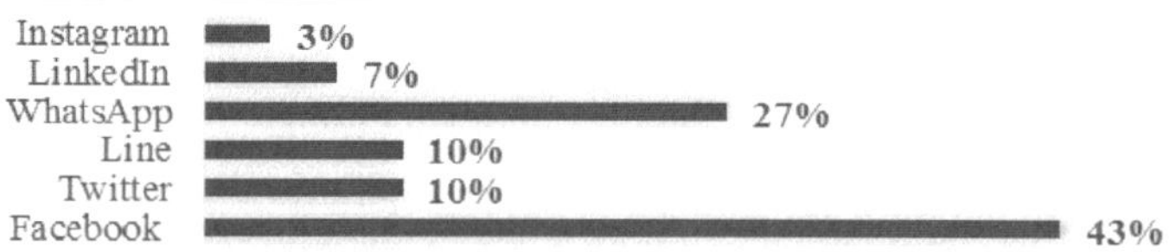

Figure 1. Types of social media sites commonly applied in teaching practices at higher institutions in Indonesia

The figure 1 above shows the majority of professional educators (43%) were regularly used Facebook rather than other social media. The second highest was WhatsApp with 27%, then followed by Twitter and Line, each applied in 10%. After that was LinkedIn (7%) and the least was Instagram 3%. The findings suggest that Facebook is the most popular social media in Indonesia, not only among students but also for the lecturers. Then, it supports Qualman's study and Falahah and Rosmala, who found that "Indonesia and United States were the country with 300 million users of Facebook, and the number of users is now creeping up" [6-7]. Therefore, it suggest that Facebook considers as a common learning tool in Indonesian education context as a learning tool. Other SMSs as LinkedIn and Instagram are ordinarily used for building relationship professionally, for business promotions, and for job seekers.

4.2. Motives Influence the integration of Social Media in Teaching Practice in Indonesian Higher Education

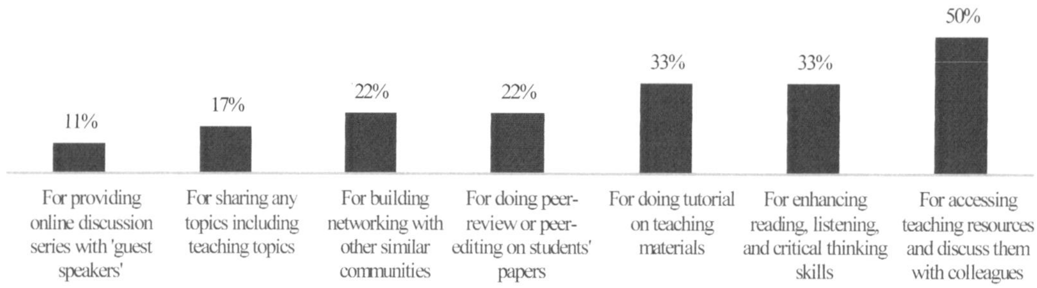

Figure 2. Motives use social media sites in teaching practices in Indonesian higher education

In figure 2, there are seven motives influenced the accommodation of Social Media Sites in teaching practices. Out of seven motive, majority (50%) of the total respondents admitted that Social Media Sites (SMSs) are a good platform for accessing teaching resources and discuss them with colleagues. Next, 33% of the respondents claimed that SMSs are good platform for enhancing reading, listening, and critical thinking skills as well as a good platform for doing tutorial on teaching materials. Then, followed by 22% of respondents that mentioned SMSs are good platform for building networking and do peer review or peer editing. Only 17% of the respondents admitted that SMSs are good platform for sharing any topics including teaching topics and 11% of total respondents agreed that SMSs are good platform for providing online discussion series with 'experts' or 'guest-speakers'. The results indicate most of professional educators' motives of using SMSs are more for professional matters than personal matters i.e. networking. These results constitutes to Srivastave [2] claims that "social media can foster the social learning since it becomes excellent tool for educators to sneak in material and ideas that students will identify with". In addition, Jenkins states that "the social media also provides the participatory learning where the social media can function as a space to allow engagement, sharing, mentoring and social interactions" [8].

4.3. Positive Values after accommodating Social Media sites in Teaching Practices

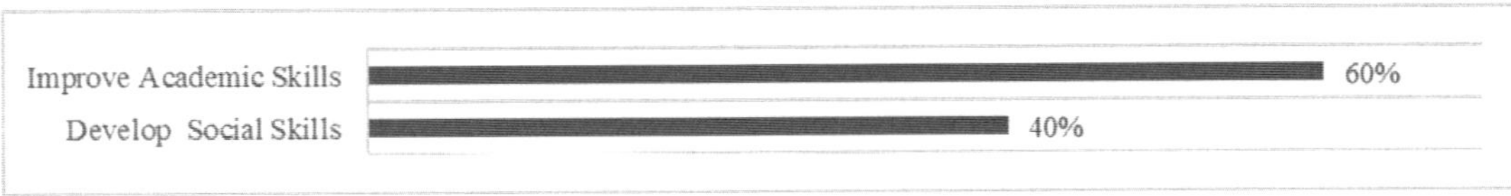

Figure 3. Positive Values after accommodating Social Media sites in Teaching Practices

As shown by Figure 3, it is found that the professional educators of two higher institutions in Indonesia had improvement in two skills, i.e. the social skills and the academic skills, after accommodating Social Media Sites (SMSs) in teaching practices. 60% or most of the respondents claimed that their academic skills were developed such as reading skills, note taking skills, listening comprehension skills, and even the presentation skills. In addition to that, 40% of the educators also improved in social skills. They could build and maintain new relationship with other peers or similar academic communities through the social media sites like Facebook, Twitter, WhatsApp and LinkedIn. These findings constitute to Gee [9] suggests that "affinity spaces instantiate participation, collaboration, distribution, and dispersion of expertise and relatedness". In other words, social media permits social cooperation and knowledge transfer in engaging interactive environment

5. Conclusions

The growing interest to integrate the advance ICT technology as social media sites in teaching practices becomes a necessity in higher institution community. The university accommodating Social Media Sites designate that the university wants to update its performance and quality, especially the professionals who works in the higher education institutions. The appliances of Social Media Sites in teaching practices highlight the two significant values of the sites to the academicians or the educators in terms of developing academic skills and social skills. Even though, the educators have different motives of implementing the social media in the teaching process, still they admit that the social media generate participatory and social leaning which useful to accommodate the real needs of the students in learning process. Thus, this study recommends that universities should allow its academicians to explore the intensity of Social Media usages in teaching practices. The study also suggests that the professional educators should use rational thinking when selecting proper social media sites which are going to be exploited in the teaching practices.

References

[1] Manca S and Ranieri M (2013) Is is a tool suitable for learning? A critical review of the literature on Facebook as a technology-enhanced learning environment *Journal of Computer Assisted Learning* **29**, 487-504

[2] Srivastava P (2012) Social Networking & Its Impact on Education-System in Contemporary Era *International Journal of Information Technology Infrastructure* **1**, 11-18

[3] Sanchez-Franco M (2010) Web CT—the quasi moderating effect of perceived affective quality on an extending technology acceptance model *Computer & Education*, **54**, 37-46

[4] Boyd D M and Ellison N B (2008) Social network sites: definition, history, and scholarship. *Journal of Computer-Mediated Communication*, **13**, 210-30.

[5] Nguyen H (2017) Indonesia is fastest growing country for internet use – report *www. rappler.com*

[6] Qualman E (2009) Statistic Show Social Media is Bigger Than You Think *www.socialnomics.net*

[7] Falahah and Rosmala D (2012) Study of Social networking Usage in Higher Education Environment in *Procedia*, **67**, 156 – 166,

[8] Jenkins H (2006). *Convergence culture: Where old and new media collide.* New York: New York University Press

[9] Gee J P (2004) Good Video Games and Good Morning *www.academiccolab.org/resources/documents/Good_Learning.pdf*

Characteristics study of the gears by the CAD/CAE

P Y Wang[1], S L Chang[2,4], B Y Lee[3], D H Nguyen[2] and C W Cao[2]

[1]Department of Mechanical Design Engineering, National Formosa University, Huwei, Taiwan
[2]Department f Power Mechanical Engineering, National Formosa University, Huwei, Taiwan
[3]Department f Mechanical and Computer-Aided Engineering, National Formosa University, Huwei, Taiwan

[4]email: changsl@nfu.edu.tw

Abstract. Gears are the most important transmission component in machines. The rapid development of the machines in industry requires a shorter time of the analysis process. In traditional, the gears are analyzed by setting up the complete mathematical model firstly, considering the profile of cutter and coordinate systems relationship between the machine and the cutter. It is a really complex and time-consuming process. Recently, the CAD/CAE software is well developed and useful in the mechanical design. In this paper, the Autodesk Inventor® software is introduced to model the spherical gears firstly, and then the models can also be transferred into ANSYS Workbench for the finite element analysis. The proposed process in this paper is helpful to the engineers to speed up the analyzing process of gears in the design stage.

1. Introduction

Gears are the most important components in the transmission system. They can be cataloged into parallel, intersected, and crossed axes transmission. The theory of gearing proposed by Litvin [1] improves the studies and application in the gearing. Many papers were then published based on the application of the proposed theory [2-5]. Based on the proposed theory, the mathematical model of the generated gear is derived according to the equation of cutter and the generating mechanism. Complex processes and developed computer program are necessary to build up the analyzing tool. Recently, the development of CAD/CAE software makes the virtual manufacture becoming possible. Some of the researches are investigated based on the CAD/CAE sources. Chang et al. [6] applied the Solidworks API to simulate the shaving process of the gears. It reveals that the method can simulate the manufacture of gears efficiently. Wang et al. [7] applied the Inventor to simulate the manufacture of spherical gears. Based on the previous study, the virtual manufacture of spherical gears and worm gear set are investigated in this paper. It reveals that the gears generated by the CAD model becomes useful and efficient for the study of gears.

In the gear transmission, the assembly error is always existed to cause the transmission error in the transmission. In traditional theory of gearing, the tooth contact analysis is applied to study the contact ellipse and transmission error causing by the manufactured and assembly errors [1, 3]. The complex equations and non-linear equation solver are necessary to obtain the contact pattern and predict the kinematic errors. It limits the further study in the gears. Now, the well-developed CAD program can be a useful tool to help the engineers and researchers to understand the characteristics of gears through the virtual manufacture.

Spherical gear is a special type of gears which could be applied in the axes with angle variation. It can be divided into concave and convex shapes which are proposed by Mitome et al. [8]. The meshing of spherical gears can be cataloged into convex-convex, convex-concave, and convex-spur types in the intersected axes transmission as Fig. 1 shows [9]. Spherical gear pairs have small kinematic errors with axis misalignment [10] and avoid the edge contact as Fig. 2 shows [9]. Theory of gearing is successfully applied to build up the mathematical models of the spherical gears [10] and other gears [11-14]. However, the complicated deriving process and programming make the analysis difficult.

Worm and worm gear set is a kind of gear popularly applied in the crossed axes. The high gear ratio and ability of self-locking are the important characteristics. The contact between the worm and worm gear is very sensitive to the errors of manufacture and assembly [15-16]. Highly non-linear equations have to be solved by the numerical methods.

Now, the complicated tooth geometry can be successfully modeled by the CAD software and transferred to the CAE software for the further study. The investigated process is reasonable and efficient in the analysis of gears. However, the results need more verification in the near future to enhance the application of CAD/CAE in gearing.

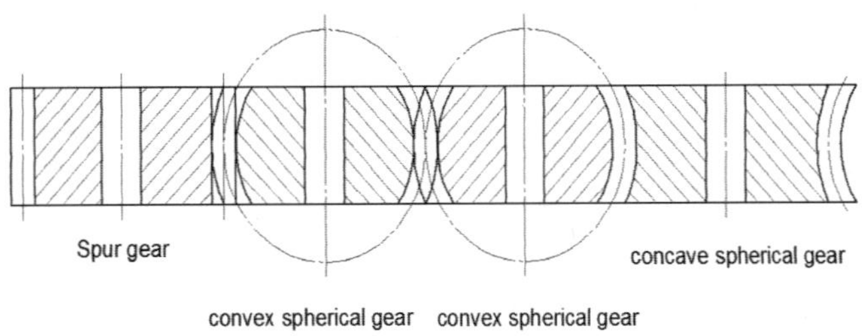

Figure 1. Different mating statuses of spherical gears [9]

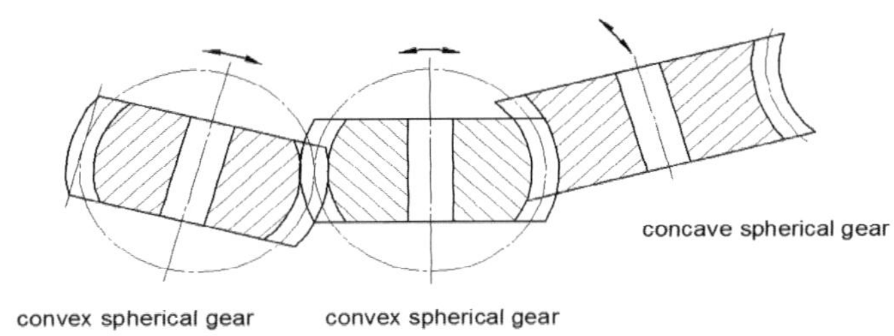

Figure 2. Spherical gear applied in the axis misalignment [9]

2. Solid models of the gears

Some kinds of CAD software are popularly applied in the geometry modeling, for example, Pro-E, Solidworks, and Inventor et al. The integration of CAD and CAE makes the software becoming powerful. The models can be transferred with each other for the further application. In this paper, the Autodesk Inventor is selected to build up the 3D solid models of the gears.

2.1 Model of the spherical gear [7]

The models of the gears can be set up by the derived equations obtained from the theory of gearing by simulating the manufacturing process. In order to avoid the complex deriving process, the simulation of the manufacturing process by CAD is applied in this paper to build the solid models of gears [7]. To build the solid model of spherical gear shown in Table 1, the processes are divided into (1) Setup the tooth profile of spur gear as the section of milling cutter; (2) Setup the gear-typed milling cutter; (3) Setup the raw material of the gear as Fig. 3 shows; (4) One tooth milling is preceded as Fig. 4 shows; (5) Complete the solid model of spherical gear.

The developed processes could be applied to generate different kinds of spherical gears with different spherical radius. Figure 5 and 6 show the convex and concave spherical gears.

Table 1. Fundamental data of the spherical gears

	Pressure Angle	*Module*	*Radius of Sphere*	*Number of Tooth*
Convex Spherical Gear	20°	2.5 mm	100 mm	14
Concave Spherical Gear	20°	2.5 mm	100 mm	14

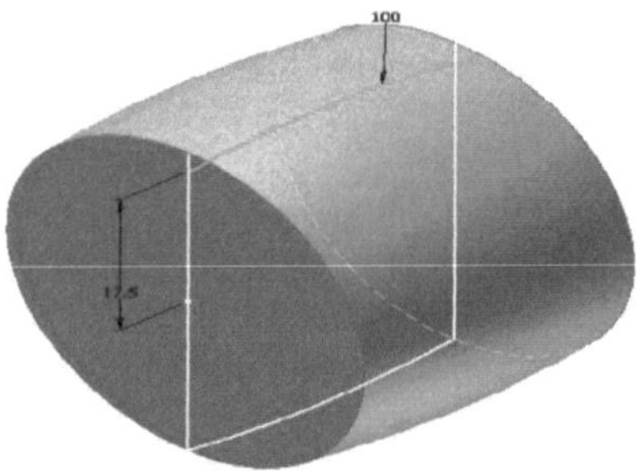

Figure 3. Raw material of a convex spherical gear

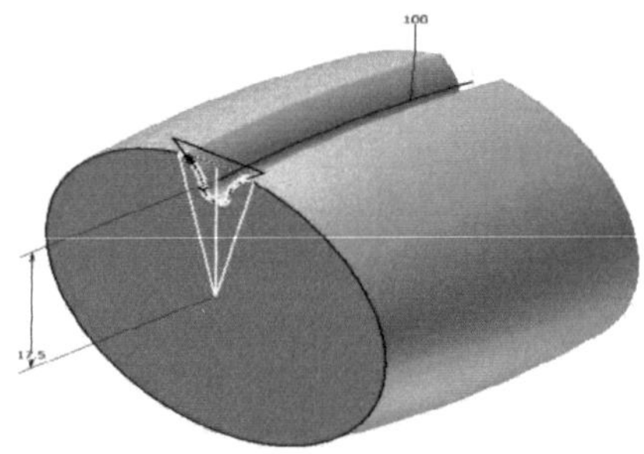

Figure 4. One tooth cut of the convex spherical gear

Figure 5. Convex spherical gear

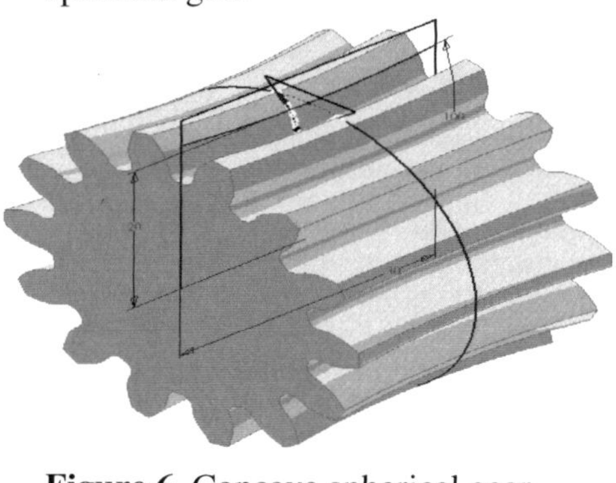

Figure 6. Concave spherical gear

2.2 Model of the Worm Gear

In the section 2.1, the models of spherical gears are simulated by the form cutter, i.e., milling cutter. There is no generation motion between the cutter and raw material. However, many kinds of gears' manufacture are based on the generation motion between the cutter and generated gears. The concept of inversed mechanism could be applied to simulate the generation process in CAD. In this section, the E-typed worm gear is illustrated to show the processes. The model of the E-typed worm could be set up firstly by rotating the cutting edge along the axis of the worm shown in Fig. 7. The worm gear is then generated by the developed worm model. When the worm gear is manufactured by the worm-typed hob cutter, the rotation angle of the worm gear is according to the rotation angle of hob cutter divided by the gear ratio of worm gear set. In developing the CAD model, the worm gear could be considered as the fixed part while the worm model is rotating and swinging along the pitch circle of the worm gear. Fig. 8 shows the assembly of the developed worm gear set by the Inventor while the parameters of the worm gear set are shown in Table 2.

Table 2. Fundamental data of the worm gear set

	Pressure Angle	*Axial Module*	*Number of Tooth*	*Pitch Diameter*
Worm	20°	3.0 mm	2	44 mm
Worm Gear	20°	3.0 mm	30	90 mm

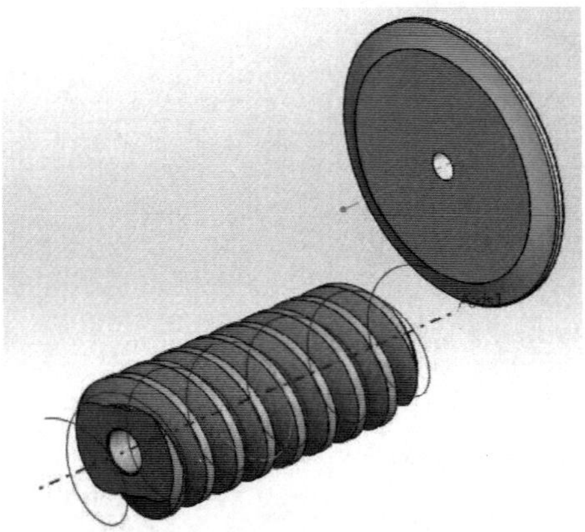

Figure 7. The model of the worm.

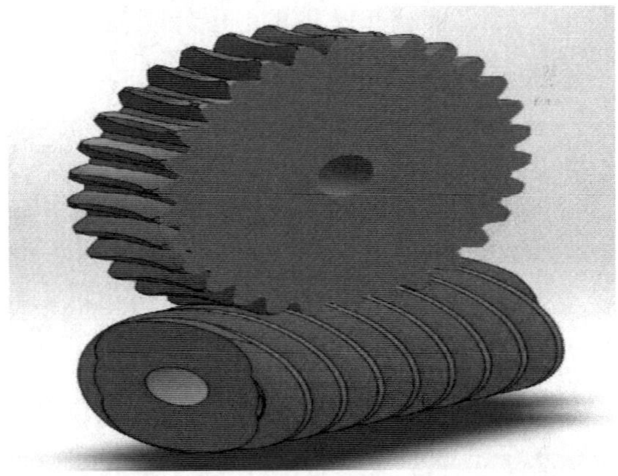

Figure 8. The assembly model of the worm gear set.

3. Stress analysis of the gears

Nowadays, the CAD/CAE software is significantly improved and can be easily applied to predict the meshing conditions of the gears, especially, gears manufactured by the complex gear machining. The complicated mathematical model might be not necessary in developing the tooth profile. The developed models of the gears by CAD could be imported into the CAE for the further study.

Wang et al. [7] studied the stress analyses for the convex-convex and convex-concave combination of spherical gears. The results revealed that the convex-concave combination can have larger contact area and lower contact stress. However, from the view of manufacture, the combination of convex-spur might be another good choice in the application. Table 3 and Figure 9 show a convex spherical gear meshed with a spur gear in parallel, while the material properties are shown in Table 4. The input torque is set as 4 N-m. The contact areas of the gears are located near the middle region shown in Figure 10.

The important advantage of the spherical gear is good at the axes angle variation. It avoids the edge contact happened in the spur and helical gears. Figure 11 shows a convex spherical gear meshed with a spur gear in 10 degrees intersected. The contact areas of the gears are shifted away the middle region shown in Figure 12. However, it reduces the worst edge contact happened in the spur gear pair and helical gear pair.

IOP Conf. Series: Materials Science and Engineering **235** (2017) 012009 doi:10.1088/1757-899X/235/1/012009

Table 3. Parameters of 18-teethed gears

	Pressure angle	Module	Radius of sphere	Number of tooth
Convex spherical gear	20°	2.5 mm	100 mm	18
Spur gear	20°	2.5 mm	--	18

Figure 9. A convex spherical gear meshed with the spur gear in parallel

Table 4. Material properties of the gears (Structure Steel)

Density	Young's modules	Possion's ratio	Tensile yield strength	Compressive yield strength
7850 kg / m³	2E+11Pa	0.3	2.5E+08 Pa	2.5E+08 Pa

(a) (b)

Figure 10. Stress analysis (a) spur Gear; (b) convex spherical gear

Figure 11. A convex spherical gear meshed with the spur gear in 10 degrees intersected

(a) (b)

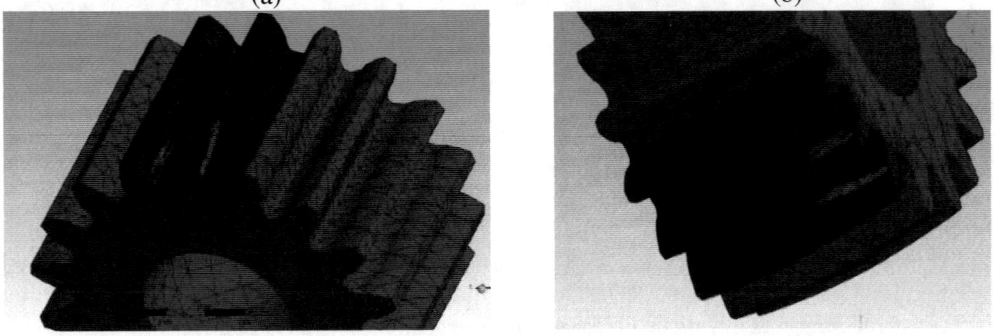

Figure 12. Stress analysis for 10 degrees intersected (a) spur Gear; (b) convex spherical gear

The mentioned process could also be applied in the study of the other types of gears. The worm gear set shown in Figure 8 is imported into the ANSYS for the stress analysis. Table 5 shows the properties of the materials while Figure 13 shows the stress obtained from the finite element analysis.

Table 5. Material properties of the gears

	Material	Density	Young's modules	Possion's ratio	Tensile yield strength	Compressive yield strength
Worm	Structural Steel	7850 kg / m³	2E+11Pa	0.3	2.5E+08 Pa	2.5E+08 Pa
Worm Gear	Copper Alloy	8300 kg / m³	1.1E+11Pa	0.34	2.8E+08 Pa	2.8E+08 Pa

4. Conclusions

The profiles of the gears are highly related to the manufacturing method and tools. The analysis of the gear meshing requires the development of the mathematical model of the profile firstly, considering the geometry of cutter and coordinate systems relationship between the machine and the cutter. It is a complex and time-consuming process for the engineer. Now, the CAD/CAE software is well developed and useful in the gear design and analysis. In this paper, the modeling and analysis processes are shown for the spherical gear and worm gear set. The models can be easily applied to predict the contact between the gears considering different gears combination and assembly. It avoids the complex process in literature and speed up the design process in the gear transmission.

(a) (b)

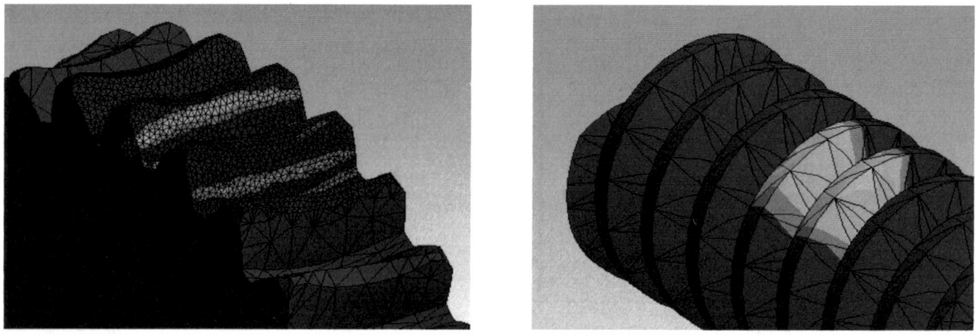

Figure 13. Stress analysis of the worm gear set (a) Worm Gear; (b) Worm

Acknowledgement
The authors are grateful to the National Science Council of the ROC for the grant. Part of this work was performed under contract No. MOST 104-2221-E-150-059.

Reference

[1] Litvin F L 1989 *Theory of Gearing* (Washington D. C.: NASA Reference Publication 1212)
[2] Chang S L and Tseng H C 2005 Design of a Novel Cutter for Manufacturing Helical Cutting Tools *Journal of Mechanical Engineering Science, Part C* **219**, pp. 395-408
[3] Liu J H, Hung C H and Chang S L 2009 Design and Manufacture of Plunge Shaving Cutter for Shaving Gears with Tooth Modifications *Journal of Advanced Manufacturing Technology* **223**, pp. 463-73
[4] Wang P Y and Fong Z H 2005 Mathematical Model of Face-Milling Spiral Bevel Gear with Modified Radial Motion (MRM) Correction *Mathematical and Computer Modelling* **41**, pp. 1307-23
[5] Litvin F L, Fuentex A, Fan Q and Handschuh R F 2002 Computerized Design, Simulation of Meshing, and Contact and Stress Analysis of Face-Milled Formate Generated Spiral Bevel Gears *Mechanism and Machine Theory* **37**, pp. 441-59
[6] Chang S L, Huang C K, Doan K H and Liu J H 2013 Serration Design and Simulation of Gear Plunge Shaving Cutter *Applied Mechanics and Materials* **376**, pp. 377-82
[7] Wang P Y, Chang S L and Cao C W 2016 CAD/CAE of Spherical Gears *Proceedings of the International Conference on Automation, Control and Robotics Engineering*, doi> 10.1145/2952744.2958871 (Kitakyushu, Japan)
[8] Mitome K, Okuda T, Ohmachi T and Yamazaki T 2000 Develop of A New Hobbing of Spherical Gear *Journal of JSME* **66**, pp. 1975-80
[9] Takashi Y 1990 The Novel Conical Gear and Spherical Gear with Variable Cross-Section Shapes *Mechanical Design* **34, No. 13** pp. 53-60 (in Japanese)
[10] Chao L C and Tsay C B 2008 Contact Characteristics of Spherical Gears *Journal of Mechanism and Machine Theory* **43** pp. 1317-31
[11] Litvin F L and Fuentes A 2004 *Gear Geometry and Applied Theory* (Cambridge University Press, Second Edition,)

IOP Conf. Series: Materials Science and Engineering **235** (2017) 012009 doi:10.1088/1757-899X/235/1/012009

[12] Tsay C B and Fong Z H 1991 Computer Simulation and Stress Analysis of Helical Gears with Pinion Circular Arc Teeth and Gear Involute Teeth *Mechanism and Machine Theory* **26** pp. 145-54

[13] Hsieh J H, Tseng H C and Chang S L 2009 A Novel Hob Cutter Design for the Manufacture of Spur-typed Cutters *Journal of Materials Processing Technology* **209** pp. 847-55

[14] Chang S L and Tseng H C 2005 Design of A Novel Cutter for Manufacturing Helical Cutting Tools *Journal of Mechanical Engineering Science, Part C* **219** pp. 395-408

[15] Chen W L and Tsay C B 2011 Mathematical Model and Tooth Surfaces of Recess Action Worm Gears with Double-Depth Teeth *Journal of Mechanism and Machine Theory* **46** pp. 1840-53

[16] Fang H S and Tsay C B 2000 Mathematical Model and Bearing Contacts of the ZN-type Worm Gear Set Cut by Oversize Hob Cutters *Mechanism and Machine Theory* **35** pp. 1689-1708.

IOP Conf. Series: Materials Science and Engineering **235** (2017) 012010 doi:10.1088/1757-899X/235/1/012010

Technical Development of Slurry Three-Dimensional Printer

Cho-Pei Jiang[1], Huang-Jan Hsu[2] and Shyh-Yuan Lee[2, 3]

[1] Department of Power Mechanical Engineering, National Formosa University, No.64, Wunhua Rd., Huwei Township, Yunlin County, 632 Taiwan
[2] School of Dentistry, National Yang-Ming University, No. 155, Sec. 2, Linong Street, Taipei, 112 Taiwan
[3] Department of Stomatology, Taipei Veterans General Hospital, No.201, Sec. 2, Shipai Rd., Beitou District, Taipei, 112 Taiwan

Corresponding e-mail: cpjiang@nfu.edu.tw

Abstract. The aim of this paper is to review the technical development of slurry three-dimensional printer (3DP) which based on photo-polymerization and constrained surface method. Basically, slurry consists of ceramic powder, resin and photo-initiator. The light engines for solidifying the photo-curable slurry can be classified as laser, liquid crystal panel (LCD), digital light processing (DLP). The slurry can be reacted and solidified by selective ray according to the reaction spectrum of photo-initiator. Ceramic powder used in this study is zirconia oxide. Experimental results show that ceramic particle size affects the viscosity of slurry severely resulting in low accuracy and the occurrence of micro crack in the layer casting procedure. Therefore, the effect of particle size on the curability and accuracy of built green part is discussed. A single dental crown is proposed to be fabricated by these three light engines as a benchmark for comparison. In addition, the cost and the limitation are compared in the aspect of dental crown fabrication. Consequently, the lowest cost is LCD-type slurry 3DP system. DLP-type slurry 3DP can produce green body with the fastest fabrication time. The volumetric error of sintered part that made by these three fabrication methods is similar because the composition of slurry is the same.

1. Introduction

The three dimensional printer (3DP) can produce parts with complex shapes using various materials. The vat photo-polymerization, one of 3DP, uses light to induce polymerization of a liquid monomer into a solid polymer. It is also known as stereolithography and divided into the top-down and the bottom-up methods as shown in Figure 1 [1]. Many 3DP developed different methods to produce 3D ceramic parts over several years. The top-down stereolithography (SL) [2], selective laser sintering (SLS) [3], fused deposition modelling (FDM) [4] and powder ink-jet printing (IJP) [5] are the typically system to fabricate the production of ceramic particles without moulds or tooling. The top-down SL can produce high accuracy green part among the others but the viscosity of slurry needs to be range between 2 and 5 Pa.s to ensure satisfactory layer recoating [6]. Furthermore, the major disadvantages of top-down SL are the high volume shrinkage ratio and unavoidable waste of slurry.

The advantages of bottom-up SL are resin savings, cost reduction and better controllable layer thickness [7]. The development sequence of light engine for SL is laser for point scan [8], optical fiber array for line scan [9], liquid crystal panel [10] and digital light projection (DLP) [11] to generate the layer pattern for layer exposure. However, the bottom-up SL based on DLP method has not been used to produce ceramic parts yet.

The aim of this paper is to compare the cost and accuracy of a single dental crown fabrication. The method used is the bottom-up SL with vary light engines consisting of laser scan, layer pattern exposure using LCD and DLP.

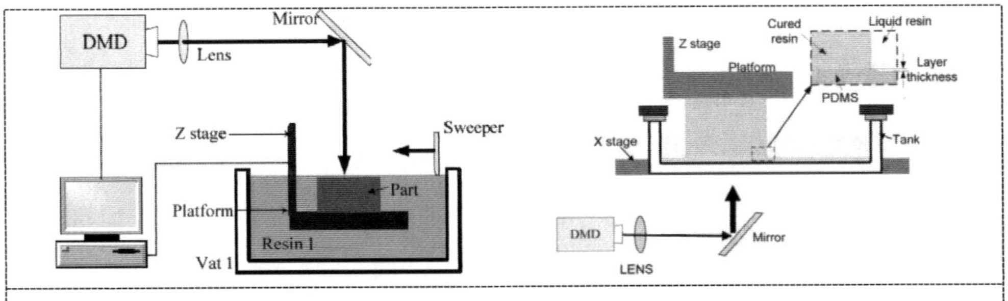

Figure 1. Two fabrication method in the sterelithography system (Left: Top-down SL system [7]; Right: Bottom-up SL system).

2. Material and methods

2.1. Slurry preparation

This study used a micro-sized zirconia-yttrium (called ZrO_2) powder with an average particle size of 1 μm. The composition is list in Table 1. The slurry consisted of four components: 70 wt% ZrO_2, 25 wt% HDDA (1,6-hexandediol diacrylate), 8.5 wt% acrylated monomer and 1.5 wt% photo-initiator (Trimethylbenzoyl diphenylphosphine oxide, TPO). The mixture slurry is then ball milled at 200 rpm for 2 hours. Finally, it is degassed in a vacuum chamber for 10 minutes to remove the air bubble that may be generated during the ball milling process. A viscometer (DV-III, ULTRA) is used to measure the viscosity because the particle size affects the viscosity of slurry and the fluidity during casting a layer. As a result, the viscosity is 3.89 Pa.s.

Table 1. The composition of zirconia-yttrium power (unit: %).

ZrO_2	Y_2O_3	Al_2O_3	SiO_2	Fe_2O_3	Na_2O	TiO_2
>94	5.2±0.2	0.25±0.05	<0.02	<0.002	<0.005	<0.002

2.2. Fabrication method

The fabrication method of this study bases on bottom-up SL system with three different light engines which are laser scanning, pattern exposure by using liquid crystal panel (LCD) and digital light processing (DLP) as shown in Figure 2. Layer pattern is scanned by laser or exposed by image projection by LCD or DLP through a transparent resin vat to cure layers that are hung upside-down on the building platform. Besides, the bottom of the resin vat is coated with a thin PDMS (polydimethylsiloxane) film to facilitate the separation of newly cured layers.

2.3. Specifications of the used system

Table 2 lists the specification of the three types of 3DP. The laser type of 3DP is Form 2 that is commercialized by Formlab Co. The resolution bases on the spot size of laser beam. The resolution of LCD depends on the pixel size as shown in Figure 3. No literature reports how to calculate the exact resolution in 3DP application using DLP. Therefore, this study proposed a general calculation method bases on the effective forming area. The DLP projector used is 720P that has a 1280 x 800 resolution. In this study, the effective forming area has a 144 wide along the X-axis. Therefore, the width (144 mm) is divided by 1280 and the exact resolution in X-axis is 0.1125. Then, the exact resolution in X-axis can be obtained by 0.1125 multiply the height resolution (800) as 0.081. As a result, the exact resolution for DLP in forming process is 113 × 800 μm. The price also lists in Table 2. The lowest price is LCD-type 3DP.

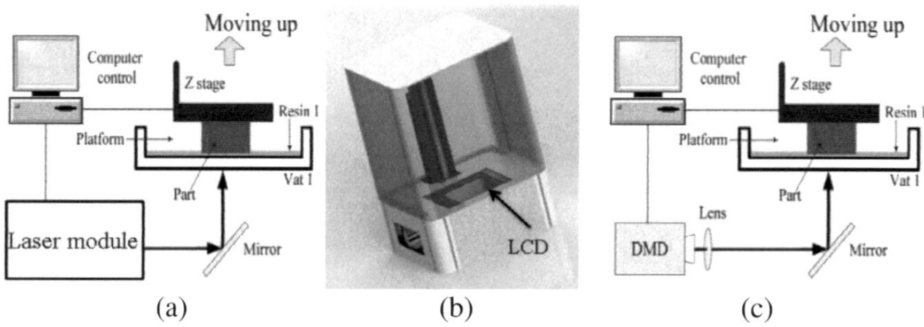

Figure 2. Fabrication method (a) laser scanning, (b) LCD light engine and (c) DLP light engine.

Table 2. The specification comparison of the three types system.

	Resolution (μm)	Average price (USD)
Formlab Form 2 (Laser scan)	Laser spot size: 140 Z-axis: 20-100	3499
LCD pattern (layer exposure)	Pixel dot: 117 × 93.6 Z-axis: 20-100	1599
DLP pattern (layer exposure)	Exact resolution: 113 × 90 Z-axis: 20-100	3488

2.4. Dental model benchmark

Single crown is an obvious aesthetic device in dental restoration device. Conventional fabrication method use CAD/CAM technique but it frequently induces the visible unavoidable surface micro-crack resulting in lower mechanical strength after sintering treatment. Therefore, this study proposes a general single crown model as a benchmark as shown in Figure 3. The dimension of this model is 10.4 × 11.5 × 6.4 mm^3. Hence, three types of 3DP will fabricate the green part of the proposed model.

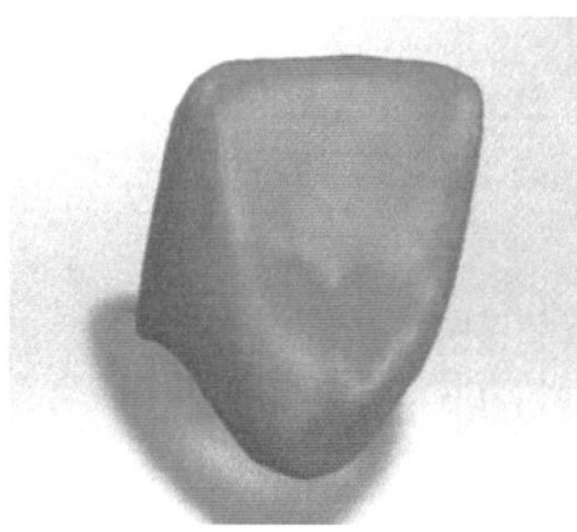

Figure 3. The proposed benchmark.

2.5. Sintering treatment

All green parts are carried out sintering treatment for obtaining zirconia dental crown with high strength and dense. Literature represents that two-stage sintering process can avoid the occurrence of micro-crack [12]. Therefore, the green parts are heated to 600 °C and maintained for 1 hour as the first stage for burnout the resin. The heating at this stage is 5 °C/min. Then, the temperature rises to 1450 °C with a heating rate of 10°C/min as the second stage. The holding time is 4 hours to allow the grain growth.

2.6. Microstructure observation and hardness measurement

The raw materials of ZrO_2 powder with fine particle size and high tetragonality could be sintered to dense at the temperature above 1400°C but crystallinity and microstructure are function of particle size and sintering parameters. Therefore, a field emission microscope (FE-SEM) is used to observe the particle size of green part and microstructure of sintered part. Furthermore, the hardness of sintered parts is measured using a Vicker test to compare with standard specimen for evaluating the hardness requirement of the 510K regulations.

2.7. Volumetric measurement

The volumetric control software is used to automatically measure the volume of each sintered benchmark. The ratio of each Form 2, LCD light engine and DLP light engine replica benchmark volumes are compared to that of the proposed benchmark CAD model.

3. Results and discussion

3.1. Green body fabrication

Figure 4(a) shows the result of green body fabrication. The layer can be seen and the thickness of each layer is 50 μm. Fabrication times for the LCD and DLP light engines are 15 minutes but for laser scan is about 28 minutes. That means laser scan needs more times to fabricate a single benchmark and depends on the numbers of fabrication piece and layer. However, fabrication time for layer exposure only depends on the number of layers. This reveals that layer exposure has higher efficacy than laser scanning. The green body is not easy to measure the dimension because the strength is low. Therefore, the volumetric measurement will be carried out by sintered part.

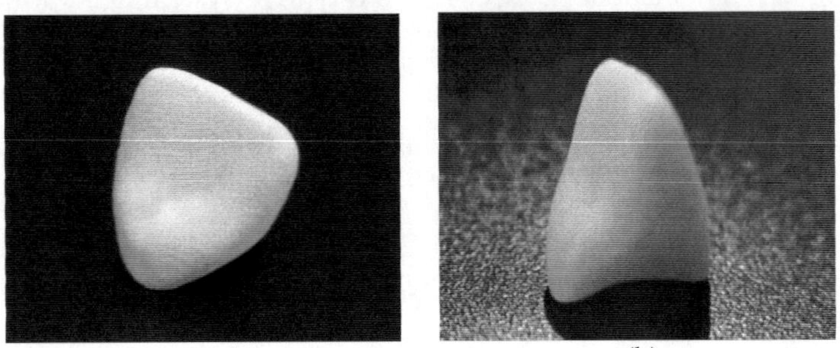

(a) (b)

Figure 4. The green body (a) and sintered part (b) of fabricated benchmark.

3.2. Sintering treatment

Figure 4 (b) represents the result of sintered part. The layer cannot be seen because the resin is burnout and shrank. Experimental results show that the volumetric shrinkage rate is about 51%. Therefore, CAD model was enlarged according to the volumetric shrinkage rate resulting in all green bodies were larger than the require benchmark. After sintering treatment, the sintered parts were volumetric measurement.

3.3. Volumetric measurement

Figure 5 shows the result of volumetric measurement. The maximum volumetric error is about 0.53 mm that occurs at the marginal line. However, the error should be less than the measured magnitude because the sintered part was painted with black due to the appearance color is white and high reflector index. The paint layer is about 0.12 mm thickness. Therefore, the exactly volumetric error is about 0.41 mm.

Because the composition of slurry is the same for the used three types of forming method, the result of volumetric error is similar. However, the fastest fabrication time for green body is the DLP type because the light intensity (58 mW/cm^2) from projection is higher than that of the LCD-type (14 mW/cm^2). As a result, the curing time for one layer of DLP-type is 2 seconds but that for LCD-type is 8 seconds.

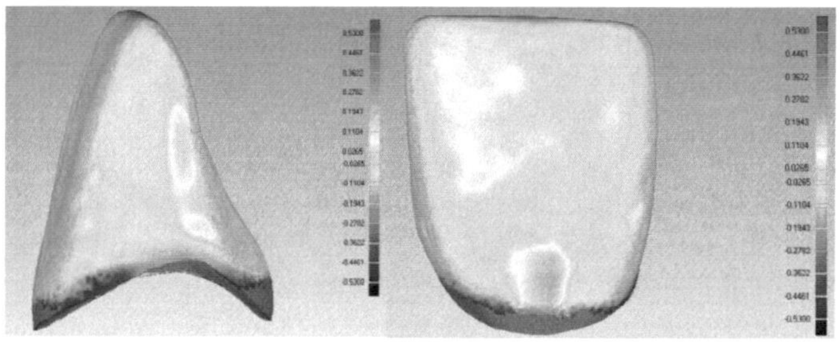

Figure 5. The volumetric tolerance of sintered part.

4. Conclusion

This study fabricates green bodies of single crown by three different types of additive manufacturing including laser scanning, LCD exposure and DLP exposure successfully. Before sintering treatment, zirconia particles are adhered by cured resin to form the green body. The volume of green body is pre-enlarged according to the volumetric shrinkage rate of sintering treatment. These three types can fabricate the green body of benchmark successfully. After sintering, the maximal volumetric error is 0.41 mm. These three methods have the similar volumetric error because the composition of slurry is the same. Comparing the equipment cost, LCD-type slurry 3DP system is the lowest. However, DLP-type slurry 3DP system can produce higher accuracy of sintered part.

Acknowledgments

The authors gratefully acknowledge the financial support from the National Science Council of Taiwan under grain no. MOST 105-2221-E-150-001 and MOST 105-2218-E-010-002.

5. References

[1] Jiang C P 2010 *Int. J Adv. Manuf. Technol.* **50**, 1003
[2] Zhou W, Li D and Wang H 2010 *Rapid Prototyping J.* **16**, 29
[3] Wilkes J, Hagedorn Y C, Meiners W and Wissenbach K 2013 *Rapid Prototyping J.* **19**, 51
[4] Turner B N, Strong R and Gold S A 2014 *Rapid Prototyping J.* **20**, 192
[5] Gans B, Duineveld P C and Schubert R S 2004 *Adv. Mater.* **16**, 203
[6] Song X, Chen Y, Lee T W, Wu S and Cheng L 2015 *J. Manuf. Process* **20**, 456
[7] Jiang C P 2011 *Rapid Prototyping J.* **17**, 148
[8] Huang Y H and Jiang C P 2003 *Int. J. Adv. Manuf. Tech.* **21**, 649
[9] Loose K, Niino T, Nakagawa T 1999 *Rapid Prototyping J.* **5**, 6
[10] Bertsch A, Zissi S, Je´ze´quel J Y, Corbel S and Andre J C 1997 *Microsys. Technol.* **3**, 42
[11] Maleksaeedi S, Wei J, Zhang K, "Bottom-up digital light projection stereolithography of zirconia ceramics," in RAPID 2014 Conference & Exposition, SME, June 9-13, 2014.
[12] Jiang C P, Hsu H J and Lee S Y 2014 *Int. J Precis. Eng. Manuf.* **15**, 2413

IOP Conf. Series: Materials Science and Engineering **235** (2017) 012011 doi:10.1088/1757-899X/235/1/012011

The Visual Representation and Acquisition of Driving Knowledge for Autonomous Vehicle

Zhaoxia Zhang[1,4], Qing Jiang[2, *], Ping Li[3], LiangTu Song[2], Rujing Wang[2], Biao Yu[4] and Tao Mei[1]

[1]University of Science and Technology of China, Heifei, China
[2]Institute of Intelligent Machines, Chinese Academy of Science, Heifei, China
[3]Anhui Xinhua College, Heifei, China
[4]Institute of Applied Technology, Hefei Institutes of Physical Science, Chinese Academy of Science, Heifei, China

*Corresponding author: jiangqing@iim.ac.cn

Abstract. In this paper, the driving knowledge base of autonomous vehicle is designed. Based on the driving knowledge modeling system, the driving knowledge of autonomous vehicle is visually acquired, managed, stored, and maintenanced, which has vital significance for creating the development platform of intelligent decision-making systems of automatic driving expert systems for autonomous vehicle.

1. Introduction

The expert systems have more applied in the field of automobile industry than others. And their development prospect is more optimistic. The technologies which are more mature, including the evaluation of automobiles overall performance, automobiles overall design, automobiles fault diagnosis, vehicle dispatch and so on. [1-5].

The expert systems of automated driving for autonomous vehicles imitate the experienced driver's action to drive. The systems applied in common vehicles which provide advice and dangerous warning to reduce driving workloads make drive easily [6-9].Expert systems applied in the unmanned vehicles, could simplify the control programs, provide reasonable decision-making schemes and control more easily [10-12].So studying the methods that acquire, manage, store and maintain driving knowledge for unmanned vehicles has vital significance for creating the development platform of intelligent decision-making systems of automatic driving expert systems for autonomous vehicle.

2. Visual acquisition and representation of driving knowledge

The driving knowledge can be divided into three categories including the traffic rules knowledge, the driving experience knowledge and driving case knowledge. And we do different processing for the three different categories of knowledge. In the driving knowledge base, we build three different kinds of knowledge bases for the three driving knowledge which are traffic rules library, driving experience library and driving case library. In order to model, express and store the knowledge, the project intends to adopt the method combined semantic web with knowledge hypergraph to achieve visualization management, storage, maintenance of the driving knowledge and facilitate the expansion of the knowledge base.

Traffic rules knowledge mainly comes from the traffic laws and regulations. And it mainly includes road traffic safety, road traffic regulations, road traffic accident treatment regulations, motor vehicle driving claims and the use rules, etc. These laws and regulations specify the various rules in effect at the time of vehicles on the road and the matters needing attention. It also uses text language for qualitative description of various traffic rules. And description of the same or similar events has different expression in different legal provisions. Nevertheless, it is a big challenge for today's unmanned vehicles to understand and identify these terms because of distinct quantitative expressions.

Figure 1 shows extraction process of the traffic rules knowledge. Collect all kinds of traffic laws and regulations in the first place. Then experts analyze these documents to extract the directly related traffic rules for driving, and select the key words that can express these rules by using the brainstorming method. The knowledge of traffic rules is transformed into the traffic rule knowledge model through the knowledge acquisition method of knowledge hypergraph. And it will use the method of concept analysis to define the concept of these words, determining the properties of vocabulary and the scope of words, making its quantitative description of the characteristics, defining the logical relationship between words, carrying on the standardization representation and finally forming the traffic rules formalization.

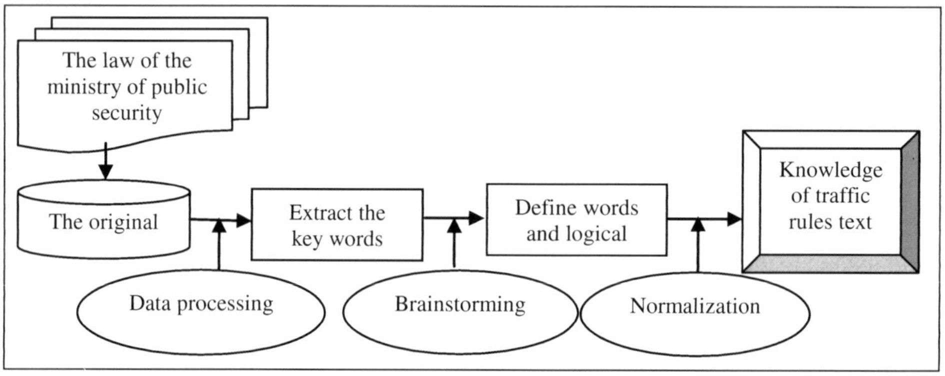

Figure 1. Traffic rule knowledge extraction process

As the knowledge of traffic rules involves more content, and the results are more complex, but the analysis of different forms will find that they have a certain structural and hierarchical. In order to make full use of the structure of the knowledge unit structure and the hierarchical structure of the knowledge unit model, the system carries on the visualization way to the knowledge modeling. The basic process of multidimensional knowledge hypergraph visual modeling mainly includes the following two parts.

- Knowledge unit division

The problem is first divided into many sub-problems, and then those sub-problems continue to subdivide until you can solve or facilitate the solution so far. With the refinement of the problem, the corresponding knowledge domain is also constantly subdivided to form knowledge units at all levels. Through the framework of the knowledge unit to describe step by step, the formation of the problem by layer decomposition.

- Knowledge unit model integration

Only through the loop, control, description, modeling assignments and other basic units to establish broad statement mapping between the evidence and the target, and there is no call other unit mode, the established knowledge unit mode is defined as the basics of unit mode. The higher-level knowledge unit model established by the basic unit model is combined with the basic knowledge unit model. The knowledge model combining the two-level knowledge unit model is defined as the three-level knowledge unit Model, and so on, for the entire knowledge base to form a multi-level knowledge unit model composed of a variety of knowledge models.

IOP Conf. Series: Materials Science and Engineering **235** (2017) 012011 doi:10.1088/1757-899X/235/1/012011

Production rules knowledge as a basic knowledge of unit cell pattern mode, part of, kind of, instance of, attribute of four kinds of basic knowledge integration relations and achieve integration with the integrated knowledge of rules of thumb, model knowledge and knowledge of the case.

After the knowledge unit is divided, the knowledge unit model is integrated from the bottom to the bottom through the guidance of the framework knowledge unit, and finally the various multi-level knowledge models needed by the users will be built. The integration process of the multi-level knowledge unit is done by the visualization of the visualization. Therefore, after the completion of multi-level knowledge model will also form a comprehensive knowledge base on the knowledge of hypergraph.

After the decomposition of knowledge unit, the integration of knowledge unit model and the integration and sharing of multi-knowledge, the process of knowledge unit editing and knowledge modeling is completed. Finally, it will realize the orderly and complex knowledge of knowledge service from quantitative calculation to qualitative reasoning effective and useful.

It has difficulties in modeling, expressing and storing if we use a single method in that the diversity of driving knowledge and the structure is complex. After analyzing the advantages and disadvantages of existing methods, the project intends to use the method combined semantic web with knowledge hypergraph to build the knowledge base of the unmanned vehicle.

The hypergraph can effectively model the multiple relationships of knowledge, express hierarchies and nonhierarchical structural relationships. The knowledge we got has different forms and hierarchies, and building the hypergraph model can model these knowledge. But there are some shortcomings in hypergraph such as difficult to store and retrieve.

The semantic web is simple, open and easy to expand, exchange and synthesize. And it contains a standard language that describes the relationship between knowledge and their relationship. Using the semantic web to express and store the knowledge base can be maintained and extended. But due to the development of the semantic web is based on the graph theory, its modeling capabilities are limited compared to the hypergraph. If we combine the two together, we can combine strengthen of the two methods to avoid shortcomings, allowing for more efficient modeling, expression and storage of the driving knowledge. Figure 2 shows the process of building a knowledge base.

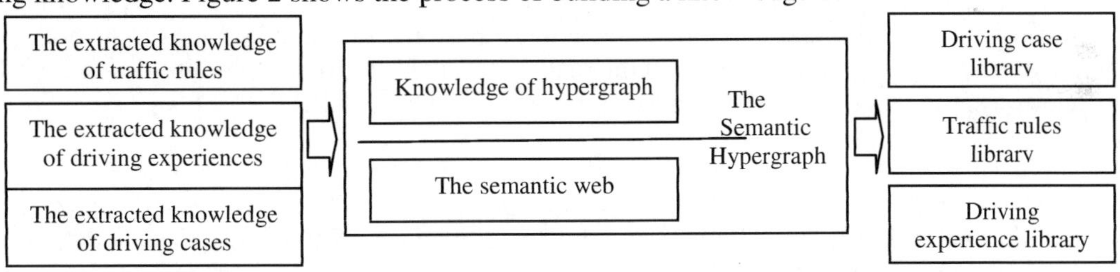

Figure 2. The build process of the construction of the knowledge base

The architecture of the proposed unmanned vehicle knowledge base is shown in Figure 3.

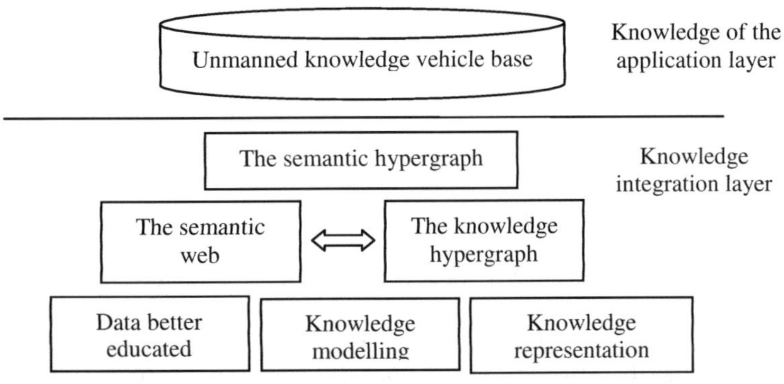

Figure 3. The unmanned vehicle knowledge base architecture

- Knowledge integration layer

The knowledge integration layer is the core of the entire framework which combined the semantic web and hypergraph. We can realize knowledge sharing, reuse and innovation through knowledge integration, and make the knowledge can be automatically analyzed, understood, and processed by computer.

- Knowledge application layer

Knowledge application layer supports the application of the unmanned vehicle and it also can provide all kinds of knowledge base required including the traffic rules library, driving experience library and driving case library.

3. The knowledge acquisition case

3.1. The extraction of knowledge

China's modern road traffic signs are divided between the main signs and the auxiliary signs, and there are 100 kinds of road traffic signs.

3.1.1. Main signs. The main sighs can be divided into the following four kinds:

- *Warning signs*: the signs are used to warn vehicles and people pay attention to dangerous place and there are 23 kinds of it. The shape of the signs is an equilateral triangle vertex angle, and its color is yellow, black and black pattern.
- *Ban signs*: it is used to ban or restrict the traffic behavior of vehicles and pedestrians, and there are 35 kinds of it. Its shape is divided into road and equilateral triangle vertex angle downward. And its color is white and red circle, black design and pattern bar, except for individual signs.
- *Indication signs*: its shape is divided into circles, rectangles and squares, and it is used to indicate the pedestrians and vehicles to get started. And there are 23 kinds of it.
- *Direction signs*: there are 20 kinds of it. And this is a sign of direction of the road, location and distance information. It is shaped by a rectangle and a square, except for the spot identification mark. And its color besides the milestone, one hundred meters of pile and the road boundary monument, the general road is the blue bottom and white design, while the highway is the green and white pattern.

3.1.2. Auxiliary signs. The auxiliary signs have five kinds of symbols, which are attached to the main sign and the role of it is to assist. And the auxiliary signs cannot be used and set up alone. Auxiliary signs according to their use are divided into the representation of time, the representation of the vehicle types, the representation of regional distance, the representation of warnings and ban reason of auxiliary sign as well as several kinds of composite auxiliary sign. Its shape is rectangular, and the color is white, black and black border. In addition, there is a kind of traffic information signs can change. It displays Information in a timely manner to the vehicle's drivers and pedestrians.

3.1.3. The geometry of a traffic sign and its meaning. In the case of the same size, the order is: the triangle, the diamond, square, regular pentagons, the hexagon, the circular and the octagon when the effect is good and it is easy to recognize. In the draft of the international security and safety standards, the regular triangle indicates a warning, a circular representation forbids and limits, and squares and rectangles are indicated. If the circular pattern has a slash, it is also forbidden.

3.1.4. The types of traffic line markings. Road traffic line can be divided into vertical lines, horizontal lines and other traffic safety lines. It has seven kinds of twenty-one species, including seventeen of them, and the rest are other traffic safety facilities. Roadway centerline, center dotted lines, solid, single center double solid line and the actual line belong to a vertical line. And the parking line, the deceleration and the crosswalk are horizontal lines.

3.2. Vocabulary, rule and knowledge meta

According to traffic rules, laws and regulations above, related vocabulary is defined, the properties and scope of the vocabulary is determined and the logical relationship between the words is defined in order to standardize them to represent the traffic rules.

After the vocabulary is defined as Figure 4, the rule is defined as Figure 5 and knowledge meta reference rule such as Figure 6.

The vocabulary unit, the rule unit and the knowledge meta form a complete hypergraph model.

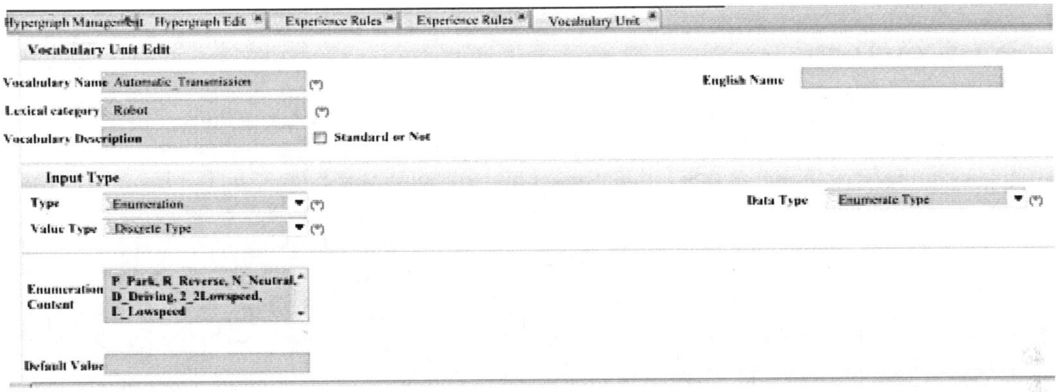

Figure 4. Vocabulary definition

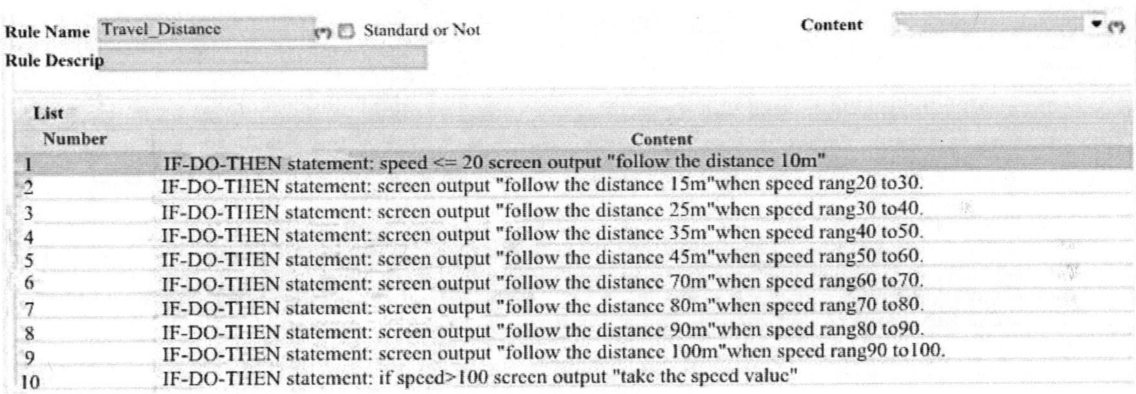

Figure 5. Regular expression

Figure 6. Knowledge super-element reference

3.3. Hypergraph knowledge base

Driving knowledge base of unmanned vehicles is established through our research and development of knowledge visualization modelling system. Firstly, we categorize driving knowledge into several sub-task, then determine the relationship of each sub-task, express the form of each sub-task node, and establish a corresponding relationship to form the hypergraph knowledge base.

Figure 7 is an example of hypergraph knowledge base of traffic behaviour.

Traffic behaviour can be two large tasks. And they are basic traffic behaviour and advanced traffic behaviour. They can also be subdivided into the following sub-task.

- Start
- The right side of passage
- Detecting point
- Lane point
- The speed limitation
- Abnormal parking
- Avoiding collision
- Safe distance

The advanced traffic behaviours include: 1. Traffic signs and traffic signs lines. 2. The recognition of traffic police commands and the traffic police command--hand signals and other traffic command signals of using tools. 3. Lane changing. 4. Returning to the original lane. 5. U-turn. 6. Automobile meeting. 7. Autonomous parking, parking in the parking lot and roadside. 8. Dynamic path planning. 9. The degradation of GPS signals: inertial navigation. 10. Emergency braking. 11. Auditory cognitive. 12. Giving way to people. 13. Using the lights. 14. Traffic of intersection. 15. Turning. 16. Driving around the islands. 17. Traffic of overpasses. 18. Traffic of railways. 19. Overpasses and railways. 20. Overflow road. 21. Integrating into the traffic flow, etc.

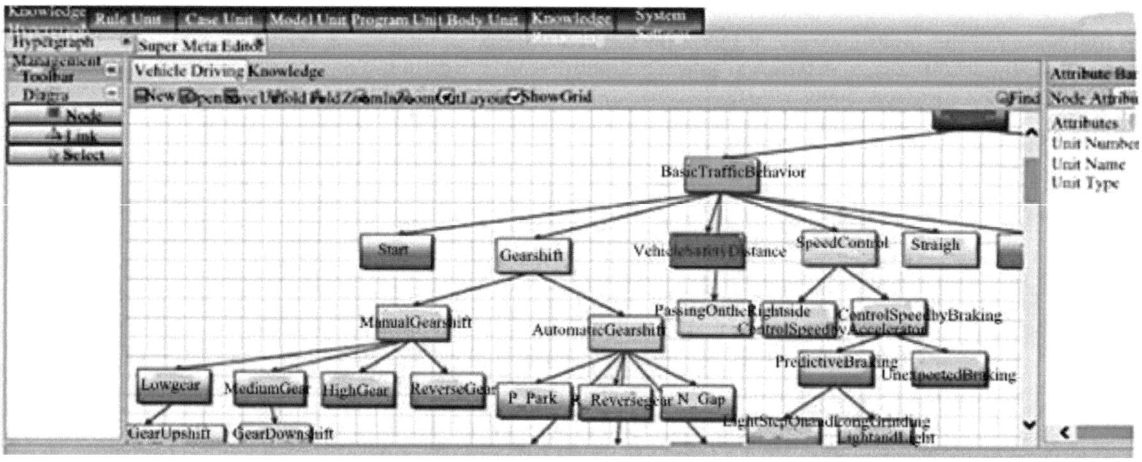

Figure 7. The hypergraph knowledge base of the traffic behaviour

4. Conclusion

According to the basic and common driving behaviors of the vehicle, the driving knowledge base of expert system for the vehicle automatic operation is established. Especially we can also use the method combined semantic web with knowledge hypergraph to achieve visualization management, storage, maintenance of the driving knowledge and facilitate the expansion of the knowledge base. After gradually improving the knowledge base, the expert system platform of automatic driving of unmanned vehicles will be finished.

Acknowledgments

Financial support from "Thirteenth Five-Year" National Key Research and Development Program of China (Grant No. 2016YFD0702002), National Natural Science Foundation of China (Grant No. 61503362, 61305111 and 91420104), Chinese Academy of Sciences students' innovation capacity training program and Natural Science Foundation of Anhui Province (Grant No. 1508085MF133) are acknowledged.

References

[1] Xiaoyun Huang, Guozhong Zhang and Fengci Zhang, Development of expert system of overall vehicle design, *Northeastern university press*, 2003.vol 42,no 1: pp 31-34
[2] Lin Hu, The design and research of software for vehicle fault diagnosis expert system, *Computer application*, 2000.3: pp 17-19
[3] Xiaojun Zheng and Shuzi Yang, Automotive engine diagnostic expert system AEDES, *Acta automatic sinica*, 1990.9.vol 16,No 5: pp 393-399
[4] Chunyi Shi, Jiaxin Wang, Xing Song and Fang Wan, A truck transport dispatch expert system, *Acta automatic sinica*, 1986.7.vol 12,no3: pp 245-251
[5] Wenyi Shan, On the establishment of traffic engineering expert system, *Highway traffic technology*, 1991.vol 8, no.2: pp 28-35
[6] Shunming Li, Huan Shen, Jianguo Mao, Jianghui Xin and Xiaodong Mou, Survey of intelligent vehicle development and its key supporting technologies, *Transducer and microsystem technologies*, 2009.vol 28, no.1: pp 1-3.
[7] Pan Zhao, Research on the control method of pilotless vehicles under the urban environment [D], University of Science and Technology of China. 2012.
[8] Rui Ling, Steering control of antonomous vehicle imitating skilled human drivers [D], Nanjing university of aeronautics and astronautics, 2012.
[9] Richard Bishop, Intelligent Vehicle Technology And Trends, 2005.
[10] Krishna Anand S, Gayathri Narayanan, Gayathri Padmanabhan , Vishnuja Sivadas U, Design of fuzzy expert system for vehicle automation, *International Journal of Engineering and Technology (IJET)*, 2012.9, vol 4, no 4: pp 238-245.
[11] Ljubo Vlacic, Michel Parent, Fumio Harashima, Intelligent Vehicle Technologies Theory and Applications, 2001
[12] Axel Niehaus, Robert F.Stengel, An expert system for automated highway driving, *IEEE Control Systems*, 1991.4: pp 53-61.

IOP Conf. Series: Materials Science and Engineering **235** (2017) 012012 doi:10.1088/1757-899X/235/1/012012

Hot Runner Mold Design of Fan Diverter Parts

D J Juan[1,2,3] **and Y L Cheng**[1,2,4]

[1]Department of Power Mechanical Engineering, National Formosa University
[2]No.64, Wenhua Rd., Huwei Township, Yunlin County 632, Taiwan (R.O.C.)
[3]juan@nfu.edu.tw

[4]10474130@gm.nfu.edu.tw

Abstract. In this study, we discuss the case of plastic parts for the production of fan steering gear shaft parts injection molding, and use POM plastic steel to produce plastic parts from traditional cold runners. Because of the parts have a hole, which need side slide. The runner produce more waste after plastic parts injection make the runner waste account for the cost is relatively high, the cost of stock preparation is relatively increased when the product quantity demanded is great. After the crushing treatment of the waste, the backfill will affect the quality, and in the crushing process, the volume generated will make the operator to withstand up to 130 dB of noise. The actual test results show that the production cycle reduce 6.25%, while the production yield increase by about 5% and material costs reduced by 2% . It can be recovered within a year, not to mention the increase of the quality and reduction the noise on the staff of the benefit is impossible to estimate.

1. Introduction

Traditional cold runner plastic injection molding, need more longer runner cooling time, the plastic loss of the runner and waste treatment, virtually reduce the efficiency and yield of molding, increase production costs. Use hot runner to conduct plastic injection production of mold, generally widely use the mode of large-scale components or multi-component molding [1], mostly still only used in 3C electronic products and high-priced parts [2], while the use environment of traditional mold is used for thin parts and multicavity molds(for example one mold eight cavity molds) such as a beverage cup. For the product of the large ratio of thickness and depth, its cooling speed shall be quick, or the front end has solidified, the latter section has not filled yet, so the plastic injection cycle shall be quick, and using hot runner can solve this problem.

If the length of general plastic pass the runner and the injection parts is relatively long, the temperature is more difficult to control, the product has the problem that the precision is difficult to control, is the opportunity to use hot runner, and the hot runner can shorten the cooling time and (multiple difference) can improve the yield. In the production process, the plastic kept at a constant temperature molten state for a long time [3], compared to the cold runner, the hot runner will not remain the waste in the cooled runner. Although the waste can

be reused after backfilling, however, there is a certain proportion of limitation of backfilling waste, cannot more than 20-25% of the part material, so as not to affect the quality is too large, and need to increase facility cost of waste treatment. The noise generated in the crushing treatment process of waste is up to 130dB, and it is an occupational injury to the health of the staff, and the heating electric energy that the hot runner need to increase itself is equal to the electric energy that the runner cooling and waste treatment need, not increase the waste of the additional energy. So the production of hot runner compared to the traditional cold runner injection molding process can reduce about 20 to 25% of the waste, while no waste generated, so do not have to deal with waste, and will not produce noise due to dispose the waste.

But even if the hot runner is used that the production would not always be perfectly conducted, it cannot be produce for the inapplicable plastic fiber reinforced material, because contain-fiber may accumulate at injection port, and make the hot runner cannot inject. However the production cost of the mould of hot runner at early stage is more than 2~4 times of the cold runner, and its structure is relatively complicated and accurate, the maintenance of mold is relative difficult [4]. But at the beginning, although the cost of hot runner is much higher than that of the cold runner, but with respect to the sophisticated parts and costly components, such as the connecting plastic components using on the car, using hot runner is the economical choice.

Currently, hot runner applications has more than 10 years , though already getting mature, but because of the substantial increase in mold production costs, the traditional industrial application of hot runner production is still small. In this paper ,we used the rotation shaft parts of traditional fan steering gear as the example as shown in Fig. 1, to investigate the niche of hot runner process production, because of its shape is similar to long shaft parts, the runner waste caused by mold arrangement account for the components material is up to 20%, in order to solve the related problem due to the waste disposal arising in the process of traditional cold runner, although the cost of hot runner mold is several times more than the traditional mold, general practitioner cannot accept, but if we have an in-depth understanding that we can know that using hot runner mold process assessment could save material fees, reduce costs of waste disposal, and avoid the damage to the staff of noise caused by crushing treatment of waste backfill and the deterioration of quality, it can also reduce the stress changes in hot and cold mold and extend the operation time. On the whole, the long run will be able to significantly reduce manufacturing costs and get better product quality, so that the degree of use of hot runner can easily accepted and promotion, which is the research motive of this article.

2. Hot runner system

Hot runner is also called micro channel, and the overall system is shown in Figure 1 [1]. It means heating in the electric coil out of the runner tube after injection each time makes the plastic in the runner free of solidification without discharging the waste in the runner when demoulding. The plastic in the runner is not solidified, so the runner is still smooth when injecting next time [5].

As shown in Figure 2 [6], when the common mold uses the cold runner, the hot plastic encounters cold mold, which recues the plastic temperature and causes pressure drop under the friction loss so that when it is injected to the mold hole, the plastic mobility has become poor and caused poorer packing and pressure holding effect and further affects the shape, appearance, dimension precision and physical property of the finished product [7]. And the waste produced must be disposed for secondary processing by the waste machine, as shown in Figure 3 [8]. If hot runner technology is applied to the mold, the temperature of hot runner is the same as that of the nozzle of injection molding machine, so it avoids the condensation of plastic on the surface of runner. Moreover, there is no runner waste when using hot runner system, so it reduces the raw material consumption and thus reduces the product cost.

In the mold cavity, the temperature and pressure are even, the stress of plastic part is small and the density is even. Under the small injection pressure and within a short molding time, the better product is molded under such system than common injection molding system.

IOP Conf. Series: Materials Science and Engineering **235** (2017) 012012 doi:10.1088/1757-899X/235/1/012012

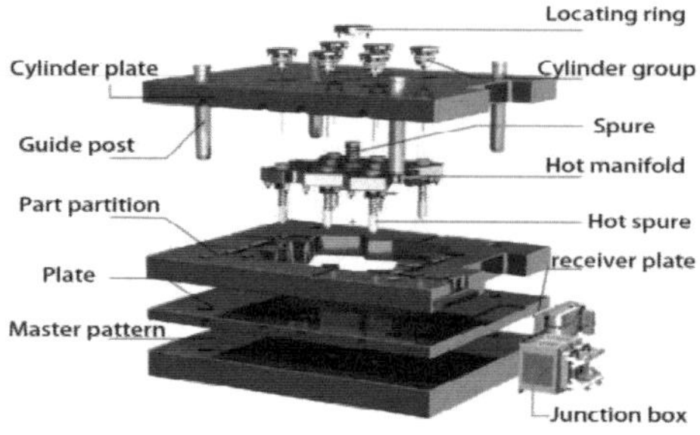

Figure 1. Hot runner explosion drawing

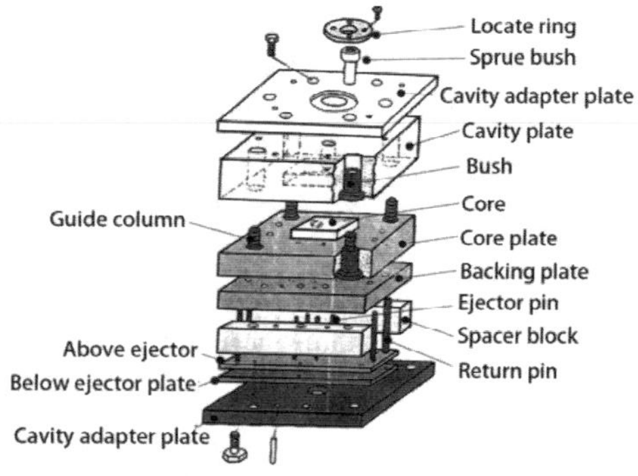

Figure 2. Traditional mold

Figure 3. Waste produced using traditional runner and waste treatment

3. Moldflow analysis of mold

Because of the parts have a hole, injection mold was limited by injection molding machine. Arrangement mode of one mold with eight holes. By using Moldex3D mold flow software, the paper analyses the packing, pressure maintaining and cooling of trial production in the mode of one mold with eight holes, predict the warping analysis of the product, observe the temperature analysis result of hot runner, optimizes the result so as to shorten the mold development and trial error process.

The material model is POM CELCON M90. The component is shown in Table 1 and the processing condition of the injection molding machine is shown in Table 2: the packing time is 1.54s, the injection pressure is 200MPa, the holding pressure is 200MPa and the cooling time is 22.4s. Through software analysis, the better packing time result is 1.564s, the holding pressure is 83.14MPa and the cooling temperature is 212.417 ℃. The mold flow analysis results are shown in Figure 4 to Figure 8 and the design case is rational, which can provide practical production of mold.

Table 1. Material component table

Items	Information
Material type	Thermoplastic
Material name	POM
Supplier	Celanese
Model	CELCON M90
T range	180-200(℃)
Mold T	60-100(℃)
Ejection time	135(℃)
Curing time	135(℃)

IOP Conf. Series: Materials Science and Engineering **235** (2017) 012012 doi:10.1088/1757-899X/235/1/012012

Table 2. Processing condition of injection molding machine

Items	Information
Filling time	1.5400(s)
Material T	190. 0(℃)
Mold T	200.00(MPa)
Injection volume	58.5029(cc)
Holding time	11.0(sec)
Max holding P	200.0(Mpa)
Ejector T	134.8(℃)
Air T	25(℃)
Cooling time	22.4(sec)
Cycle time	40.0(sec)

Figure 4. Arrangement mode of one mold with eight holes

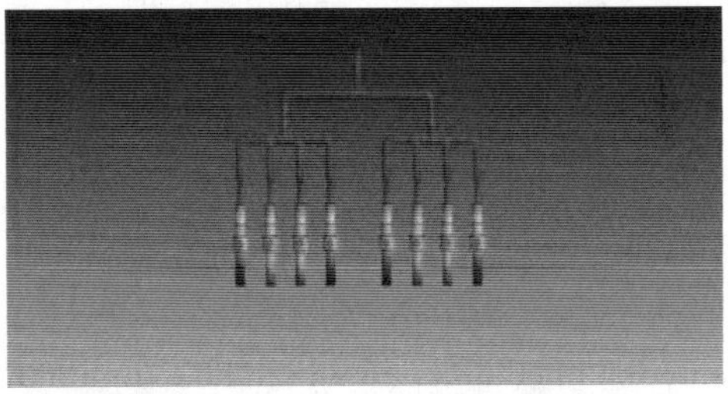

Figure 5. Packing analysis result

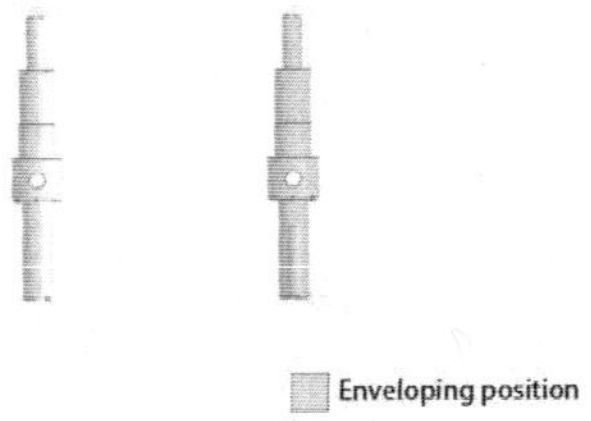

Figure 6. Packing analysis & enveloping position

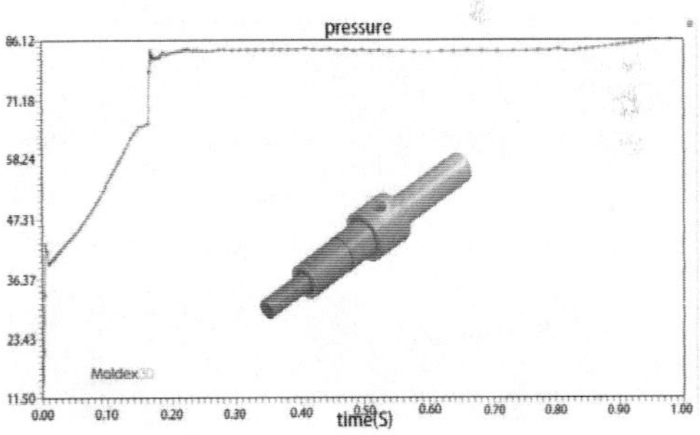

Figure 7. Pressure maintaining analysis

Figure 8. Cooling analysis

IOP Conf. Series: Materials Science and Engineering **235** (2017) 012012 doi:10.1088/1757-899X/235/1/012012

4. Mold design and manufacturing

As shown in Figure 9 to Figure 12, the hot runner system diagram of one mold with eight holes, the same arrangement mode as the traditional mold is mapped by referring to the analytic result of mold flow software and using SolidWorks plotting software. After mapping, the mold flow analysis such as packing, pressure maintaining and cooling of runner should be carried out by using Moldex3D software.

Figure 9. Male mold

Figure 10. Master pattern of mold

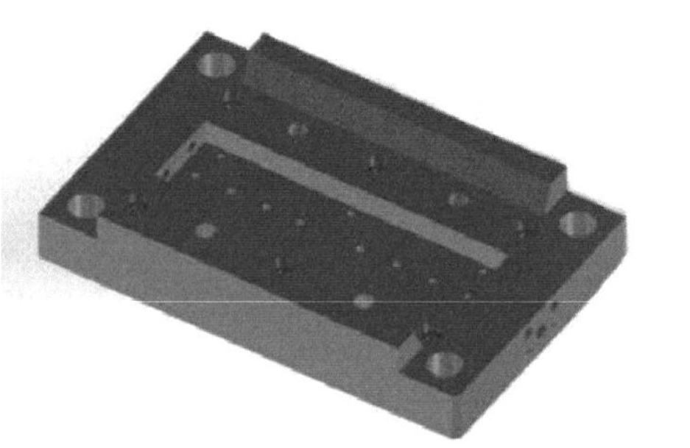

Figure 11. Prt partition

Figure 12. Ht runner plate

Taking the analysed parameters as reference for mold development can reduce the trial error process. The manufactured result is shown in Figure 13. to Figure 15.

Figure 13. Noumenon of mold

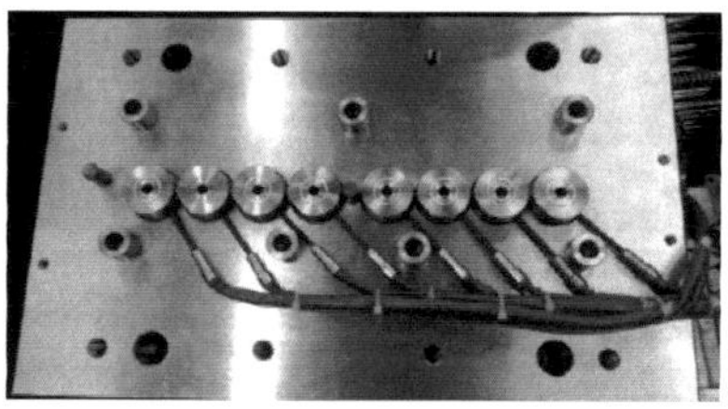

Figure 14. Part partition

Figure 15. Hot runner gate

5. Analysis and comparison

Using the Injection rate, holding pressure and holding time as experimental factor, each factor adopted 3 levels, used statistical analysis software minitab to conduct Taguchi statistics could obtain L_9 (3^4) orthogonal table, the combination of each experimental parameters repeated 3 times and conducted 27 experiments, and the injection and measurement results are as shown in Table 3.

Figure 16. injection Experiment

And then according to the injection experiment of plastic parts of experimental combination shown in Figure 16, due to the shape of the shaft is step-shaped and its number is large, it can`t be detected by general projector, so chose flat head gauge at 0.01mm precision to measure the deflection on CNC , firstly using standard round bar with HSS TOOL μ-class as the calibration benchmark of the base and the fixture, the deflection is in the 1μm, after calibration, the measurement component of the Run Out value is shown in Figure 17 and Figure 18. And Run Out value of the shaft parts needed to measure is the smaller the better, is a smaller-the-better type characteristic.

Figure 17&18. Instrument calibration

After experiment, get the SN ratios. And the optimal parameters of injection rate, injection pressure, holding pressure and holding time are 68mm/s, $58kg/cm^2$, 35% and 3.6 second respectively.

Table 3. Experiment parameters and SN ratios

No.	Injection rate	Injection pressure	Holding pressure	Holding time	Y1	Y2	Y3	Y4	Y5	Y6	Y7	Y8	SN
1					0.12	0.23	0.67	0.22	0.43	0.26	0.19	0.57	
2	60	58	25	3.0	0.17	0.2	0.28	0.18	0.3	0.12	0.34	0.14	9.99
3					0.16	0.18	0.19	0.35	0.16	0.27	0.64	0.29	
4					0.24	0.09	0.07	0.52	0.19	0.17	0.19	0.29	
5	60	63	30	3.6	0.36	0.28	0.29	0.45	0.02	0.28	0.04	0.26	11.71
6					0.22	0.25	0.36	0.03	0.18	0.32	0.25	0.11	
7					0.35	0.16	0.37	0.34	0.33	0.25	0.21	0.08	
8	60	68	35	4.2	0.07	0.48	0.18	0.36	0.23	0.17	0.55	0.33	10.85
9					0.32	0.2	0.21	0.15	0.12	0.38	0.17	0.25	
10					0.51	0.29	0.49	0.03	0.15	0.15	0.29	0.27	
11	64	58	30	4.2	0.57	0.54	0.2	0.14	0.17	0.31	0.24	0.4	10.63
12					0.01	0.15	0.25	0.17	0.31	0.23	0.06	0.06	
13					0.25	0.2	0.53	0.51	0.39	0.46	0.57	0.24	
14	64	63	35	3.0	0.34	0.07	0.05	0.37	0.3	0.19	0.13	0.22	9.13
15					0.09	0.16	0.52	0.2	0.48	0.22	0.53	0.43	
16					0.45	0.63	0.05	0.12	0.57	0.23	0.29	0.31	
17	64	68	25	3.6	0.56	0.64	0.47	0.18	0.25	0.34	0.3	0.08	9.08
18					0.37	0.25	0.01	0.12	0.02	0.24	0.3	0.42	
19					0.06	0.02	0.01	0.05	0.28	0.07	0.15	0.28	
20	68	58	35	3.6	0.38	0.4	0.13	0.28	0.1	0.29	0.14	0.27	13.12
21					0.22	0.12	0.2	0.14	0.39	0.3	0.01	0.17	
22					0.23	0.4	0.34	0.07	0.08	0.48	0.14	0.22	
23	68	63	25	4.2	0.45	0.25	0.05	0.08	0.28	0.14	0.16	0.36	10.77
24					0.57	0.01	0.23	0.27	0.37	0.12	0.24	0.43	
25					0.02	0.22	0.46	0.52	0.03	0.3	0.1	0.44	
26	68	68	30	3.0	0.45	0.22	0.46	0.52	0.07	0.1	0.22	0.29	10.41
27					0.08	0.01	0.18	0.14	0.19	0.26	0.26	0.5	

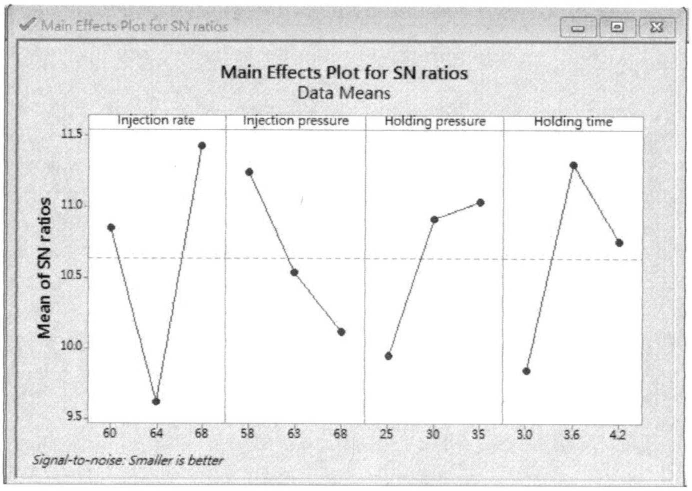

Figure 19. SN ratios

The injection molding of product using traditional cold runner needs longer gate cooling time, plastic loss of gate and waste treatment, which reduces the productivity and increases the cost. The waste produced in the gate can be recycled and mixed with new raw material for secondary injection molding and completing the consecutive processing work, but waste backfilling also affects the product quality and the maximum waste packing ratio cannot exceed 20% to avoid affecting the quality of plastic parts. Meanwhile, the noise produced in the treatment process is as high as 130 decibel, which affects operator's hearing seriously. In addition, the waste grinding process also needs additional energy and thus forms energy waste. Therefore, compared with the injection molding with traditional cold runner, the production using hot runner can reduce the waste by 20%-25% and thus reduces the noise of disposing the waste. The different is shown in Figure 19 and Figure 20.

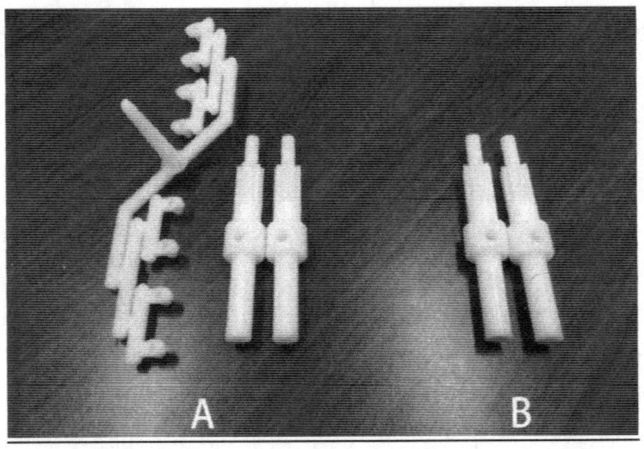

Figure 20. (a) & (b) Parts and wastes produced by using traditional cold runner

6. Conclusion

The application and production of hot runner mold centers on large plastic parts or electronics parts with high unit price. The mold manufacturing cost is over two or three times than that in traditional form, so it is seldom applied to traditional home appliances with meager profit.

The manufacturing with hot runner system mold can increase the annual yield. The single production cycle of parts injected by the hot runner is shortened from the original 85s to 80s, with 6.25% shortened than the original production cycle. If one day is calculated as eight working hours, one mold with eight holes can increase 170 parts, the annual increased yield reaches over 45,000, and the increased capacity can reach over US$3,900. Moreover, there is no waste produced and no need waste recycling, which enhances the production yield rate from 99.9% to 99.95% ,while the production yield increase by about 5% and material costs reduced by 2% and reduces the noise greatly that operators don't have to bear the noise as high as 130 decibel.

References

[1] http://anntong.com.tw/Hospital/Japan/Reclaim3-2.htm (Ann Tong Ind.Co.,LTD)
[2] Meng-Yu Jiang 2005 *Implementation of Micro Hot Runner System in Thermoplastic Injection Molding Multiple Micro Component* (National Yunlin University of Science and Technology)
[3] Yong-Yan Zhang 2013 *Plastics Mold Design* (Chuan Hwa Book Co.,LTD)
[4] Zhen-Li Xu 1987 *Use of Hot Runner*, research report (AnnTong Co.,LTD)
[5] Guo-Ying Wu 2001 *Design and analysis of hot runner* (Metal industries research & development centre)
[6] http://www.chinabaike.com/t/30170/2015/0921/3342263.html(website)
[7] Chen Hsong Monthly Publication 2012 *CAE New injection simulation analyse in hot runner and Conformal Cooling* (CHEN HSONG MACHINERY TAIWAN CO. ,LTD)
[8] http://www.cens.com/cens/html/en/supplier/supplier_home_3880.html (SHINN YUH PLASTIC CO., LTD.)

IOP Conf. Series: Materials Science and Engineering **235** (2017) 012013 doi:10.1088/1757-899X/235/1/012013

Series Hybrid Electric Vehicle Power System Optimization Based on Genetic Algorithm

Tianjun Zhu[1,3], Bin Li[2], Changfu Zong[1] and Yang Wu[3]

[1] Department of Electronic Information and Electrical Engineering, ZhaoQing University, ZhaoQing, 526061, China
[2]CONCAVE Research Center, Department of Mechanical and Industrial Engineering, Concordia University, Montreal, Quebec, H3G 1M8, Canada
[3]College of Mechanical and Equipment Engineering, Hebei University of Engineering, Handan, 056038, China

happy.adam2012@hotmail.com

Abstract. Hybrid electric vehicles (HEV), compared with conventional vehicles, have complex structures and more component parameters. If variables optimization designs are carried on all these parameters, it will increase the difficulty and the convergence of algorithm program, so this paper chooses the parameters which has a major influence on the vehicle fuel consumption to make it all work at maximum efficiency. First, HEV powertrain components modelling are built. Second, taking a tandem hybrid structure as an example, genetic algorithm is used in this paper to optimize fuel consumption and emissions. Simulation results in ADVISOR verify the feasibility of the proposed genetic optimization algorithm.

1. Introduction

Hybrid electric vehicles can save fuel and reduce environmental pollution [1, 2]. Compared with conventional vehicles, HEV have a huge advantage and market prospect. However, because of the complicated hybrid structure, the relationships among some vehicle components are not clear. Although a variety of control strategies and energy management systems have been developed and applied to HEV, there are still much work to do for further reducing the fuel consumption and emissions of vehicles. The interactions among different vehicle components need to be further investigated to provide theoretical foundations for a more practical HEV powertrain controller design.
In recent years, many studies from academy and industry have been carried out on how to reduce emissions and save fuel using HEV with various control algorithms and strategies. The intelligent control algorithm, such as genetic algorithm (GA), particle swarm optimization (PSO) algorithm, dynamic programming algorithm and artificial neural network, have gained much attention [3]. For example, Yan [4], without sacrificing the premise of vehicle dynamic performance, employed genetic algorithm to optimize the HEV parameters configuration with improved fuel economy and reduced emissions.

This paper aims to optimize the component size to make it work in a more efficient manner so as to reduce fuel consumption and emissions. Taking the series hybrid electric vehicle as example, based on the modelling of the most important parts, the relationship between components and their optimal working range can be found, particularly the ratio between the generator speed and gear in "optimization" zone of the engine. The organization of the rest of this paper is as follows. First, the basic vehicle configuration and parameters are given.; then the genetic algorithm and Control Strategy Optimization part of ADVISOR are used to set the parameter range; Furthermore, simulation are conducted to find out the most reasonable size of components and parameters; Finally, compared with the original size, the optimized components size are verified to achieve the better vehicle performance.

2. HEV powertrain system

2.1. Hybrid electric vehicle components

Series hybrid electric vehicle components are shown in the Figure 1.

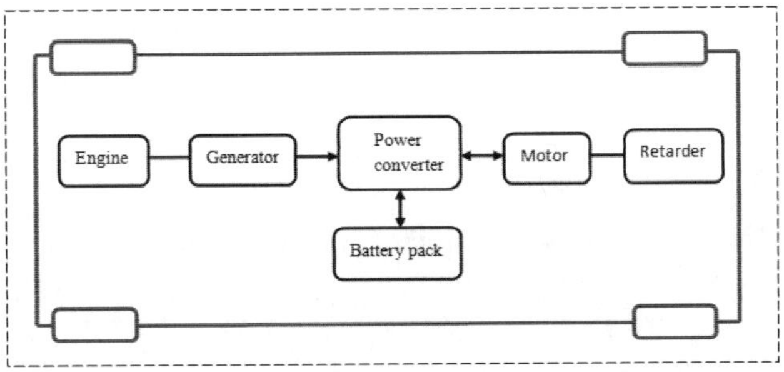

Figure 1. The hybrid powertrain system components

Series hybrid electric vehicle parameters in detail are shown in the Table 1.

Table 1. Parameters of series hybrid electric vehicle.

Component	Parameter	Value	Component	parameter	standard	name	parameter	standard
Engine	CC	1.6L	lithium battery	mold	lithium battery	generator	maximum power	75Kw
	maximum power	41Kw		The battery block	45~50		maximum speed	8000r/min
	Maximum torque	160Nm		single marker	26Ah		Maximum torque	120Nm
Motor	maximum power	80Kw		energy density	105W·h/kg	vehicle parameter	vehicle weight	1136kg
	maximum speed	10000r/min		power density	1000w/kg		windward area	2.0m^2
	Maximum torque	273Nm		total voltage	360~420v		drag coefficient	0.335
	Mold	asynchronous motor					wheel base	2.6m

3. Optimization of series hybrid electric vehicle components

3.1. Optimization objective

The goals of HEV optimization are to reduce fuel consumption and emissions such as HC, NOx, CO, etc as much as possible under the premise of guaranteeing the vehicle dynamic performance [5]. These three kinds of pollutants: HC, NOx, and CO with the same unit of measurement, are considered as the terms of optimization goals. According to the requirements, equivalent fuel consumption per hundred kilometres is taken as another term of optimization goals. Using ADVISOR software is used in this paper for car simulation and optimization process. The optimization goal in this study is defined as follows:

$$\begin{cases} \min f(x) = 0.7Fuel(X) + 0.1CO(X) + 0.1HC(X) + 0.1NO_X(X) \\ X \in \Omega \\ st \quad g_i(x) \ge 0 \qquad j = 1,2..m \\ \quad x_i^{\min} \le x_i \le x_i^{\max} \quad i = 1,2..m \end{cases} \tag{1}$$

where X is the vector of parameters to be optimized; Ω is the feasible solution space; x_i is the parameter to be optimized; n is the number of parameters; $g_i(x) \ge 0$ is constraint conditions which describe the vehicle's performance requirements such as maximum speed, acceleration time and climbing capacity. $[Fuel(X), O(X), HC(X), NO_X, COST(X)]$ are multiple objective functions, followed by fuel consumption per hundred kilometres and carbon tariff. It is refined as follows:

The design variables: $X^T = \{x_1, x_2.x_3, x_4\}^T =$ {Engine power, generator power, motor power, battery power} The objective function: $f(x) = \{f_1(x), f_2(x), f_3(x)\} =$ {Fuel economy, emissions, car costs}

The constraint condition: $g(x) = \{g_1(x), g_2(x), g_3(x)\} =$ {Top speed, climbing, acceleration} The Boundary constraints: $x_{min} \le x_i \le x_{max}$

Table 2. The optimization target and weight coefficient

Measurement	Target	Unit	Weight coefficient
Fuel	6	L/100km	0.7
HC	0.6	g/km	0.1
CO	1	g/km	0.1
NO_x	0.27	g/km	0.1

Table 2 show the optimization targets and the weight coefficients in the optimization objective function.Using genetic algorithm, it sets the evolutionary population size as 100, biggest evolution algebra as 50, the mutation probability as 0.1, crossover probability as 0.9, and other relevant genetic parameters use the default values. The optimization algorithm is based on simulation results generated in ADVISOR software to analyse the evaluation of the vehicle performance. Figure 2 shows the running condition for simulations, which is composed of three types of working conditions. Table 3 shows the optimized design variables and control variables including three component parameters and four control strategies, and the constraint condition is the original vehicle dynamic performance requirements. The following are the design variables, using the control strategy optimization of ADVISOR as simulation tools [6].Table 3 shows the minimum motor power and maximum battery capacity. The most economic mode is when it is in pure electricity mode and the dominant factor is the determination of upper and lower limits of power elements. Based on the simulation result to figure out target function and constraint and then collect the data of fuel consumption and emission.

Table 3. Mixed power system optimization parameters

Design variable	Engine power	Motor power	Battery capacity	Highest SOC	Lower SOC	Max power command	Min power command
default	41kW	80 kW	26A·h	0.8	0.2	30Kw	20kw
lower limit	25 kW	38 kW	13A·h	0.7	0.2	25 kW	5 kW
upper limit	53kW	110Kw	39A·h	0.9	0.5	41Kw	20kW

3.2. Driving cycles

As road traffic environment is complicated and variable, some research only chose single road condition as the selection of driving cycles. Although some good results have been obtained, single driving cycle cannot cover the real driving patterns. The complicated working conditions are selected in the simulation. As the simulation target, to be exact, the working condition is a combination of urban roads and highways about 64 kilometres [7].

3.3. Simulation and results analysis

In this part, three driving cycles are combined into a single loop combination as the mixed operating mode, as shown in Figure 2.

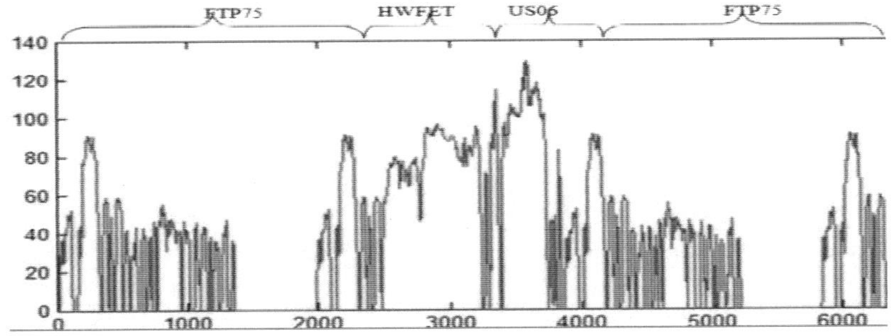

Figure 2. Combined city and highway driving cycles

Before optimization, the vehicle model is built to meet the requirements of standard dynamic performance [8].

1) 0-96.6km/h: 10.5s; 2) 64.6-96.6km/h: 5.6s; 3) 0-137km/h: 24.6s; 4) At 96.6 km/h speed climbed the slope of 25%. Figure 3-Figure 5 respectively shows that the values of three emssions HC, NO_x, CO with the change of the number of iterations.

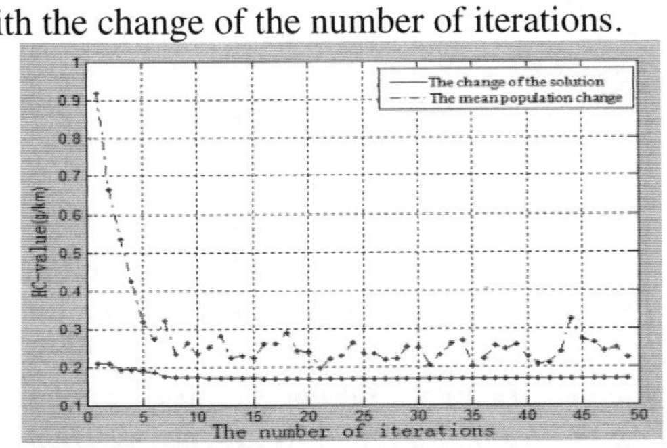

Figure 3. *HC* Emission changing curve

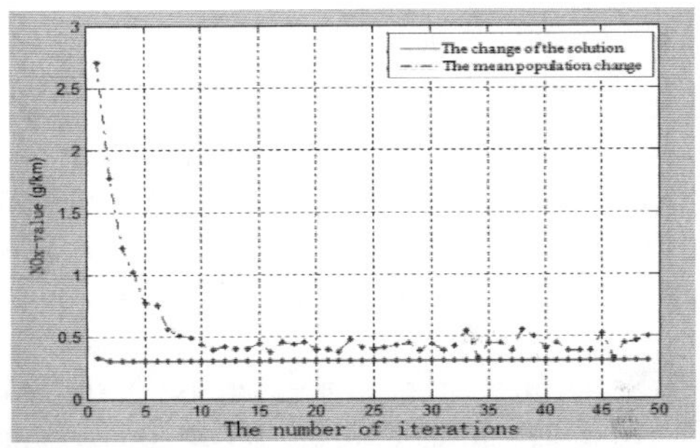

Figure 4. NO_x Emission changing cur

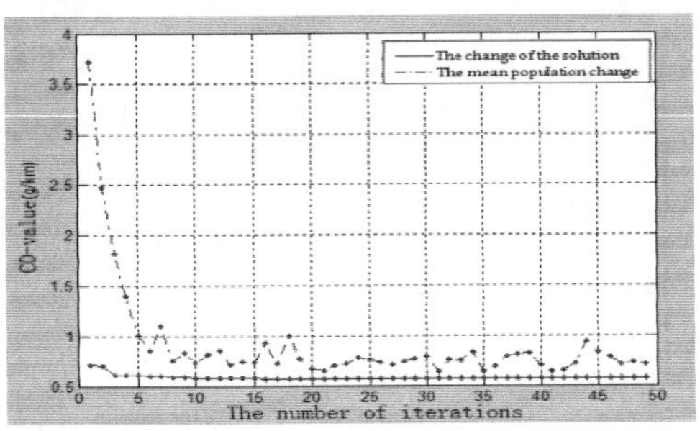

Figure 5. *CO* Emission changing curve

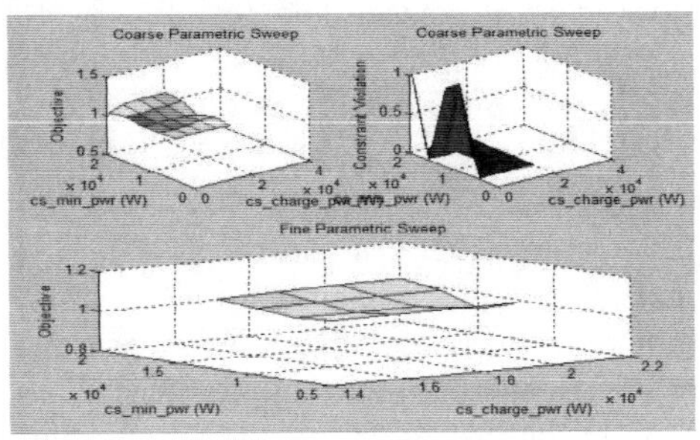

Figure 6. *CO* Emission changing curve

From Figures 3-5, it can be seen that the initial values of *HC, NO_X, CO* are 0.21 g/km, 0.35 g/km and 0.70 g/km, respectively. With the increase of the number of iterations, the optimal values of three kinds of emissions decrease gradually and finally tend to be stable with optimal values 0.17 g/km, 0.30 g/km and 0.60 g/km or so. In terms of pollutant emissions, *CO* emissions obtain the largest decline, followed by HC. Although NO_x increases slightly, the overall emissions decrease, which is beneficial to the environment. The sum of the three kinds of pollutants emissions is shown in Figure 6. From the results of Figure 6, for coarse parametric sweep which is rough scanning parameters, the sum of the three pollutants are between 1 g/km and 1.5 g/km, tending to be 1 g/km, while for fine parametric sweep which is precise scanning parameters, the result is more accurate than before, and the interval narrows down to 1 g/km ~ 1.2 g/km. Based on the simulation results for 5 groups from ADVISOR, the fuel consumption and emission values before and after optimization are listed in Table 4. The simulation results demonstrate that fuel consumption and the emission have been significantly reduced with the proposed genetic algorithm. The optimal simulation is also helpful in optimizing the vehicle components. The simulation solutions for the following 5 groups are shown as the Table 5.

Table 4. The optimization results

No.	*Fuel*/ (L/100km)	*HC*/(g/km)	CO/(g/km)	NO_X /(g/km)
1	5.6	0.145	0.698	0.280
2	5.8	0.138	0.592	0.275
3	5.8	0.145	0.680	0.292
4	5.7	0.148	0.715	0.299
5	5.9	0.168	0.827	0.315

Table 5. Optimal design variables and performance values

Sol.	Engine power	Motor power	Battery capacity	Highest SOC	Lower SOC	Max power command	Min power command	0~96 km/h	40~96 km/h	0~137 km/h	Climbing ability of 96.6 km/h
1	31 kW	88kW	30.2A·h	0.79	0.30	33.1 kW	13.7 kW	7.7s	3.6s	16.4 s	26.0%
2	28 kW	92kW	29.8A·h	0.81	0.35	36.6kW	13.3kW	7.3s	3.3s	15s	27.6%
3	29.2kW	96kW	28.9A·h	0.82	0.32	32.2 kW	9.8 kW	7.7s	3.6 s	16.5s	25.9%
4	30 kW	90kW	27.2A·h	0.8	0.30	33 kW	5.6 kW	7.2s	3.3s	15s	27.2%
5	40 kW	94kW	26.5A·h	0.84	0.30	32kW	12.4 kW	7.5s	3.5 s	16s	27.8%

According to Table 4, the emissions obtained by genetic algorithm and the ADVISOR simulation are in the same order of magnitude despite of somewhat gap. And this shows that the proposed optimization algorithm for component parameters and fuel consumption is feasible. And Table 5 shows the optimized design variables and performance values. For example, comparing Solution 1 with Solution 2, although the engine power is reduced, vehicle's acceleration and climbing ability are improved by increasing the motor power and the capacity of the battery. Solution 4 shows that the emission is not as satisfactory as the solution 3 even when the number of batteries and the power of the motor are increased and thereby the cost is increased by 5%From the optimization results, it is clear that the good matching of components could have positive effect on the vehicle performance and emission of hybrid vehicle.

4. Conclusions

In this paper, genetic algorithm is used to optimize the components and fuel consumption of series hybrid vehicle. Simulation results show that fuel economy and emissions can be optimized simultaneously. By comparing the selection of appropriate component parameters through analyzing the five values obtained by optimization, this research narrows the optimization area of the region to maximum optimization value.Due to the inadequate number of iterations of algorithm and the selection of fitness function, there are some difference between the results and that of the ADVISOR. The further research will focus on increasing the number of iterations and adjustment of the corresponding algorithm parameters, such as the number of lithium batteries, the engine unit transmission ratio, and the motor with high rotating speed and smaller peak power, in order to improve efficiency and reduce the weight of the motor.

In general, the genetic algorithm can make an easier and more effective search to the optimal solution of hybrid cars system parameters, and call to optimize parameters can improve the matching of components and fuel economy, thus improving the overall performance of the vehicle, genetic which also verified the effectiveness of the hybrid system optimization problems.

Acknowledgments

The authors would like to acknowledge the financial support of Collaborative Innovation and Platform Environment Construction Project of Guangdong Province (2015A050502053) , Hebei science and technology plan project(17394501D), Science and technology research project of Hebei higher education (ZD2017213) Hebei provincial High Level Talents Foundation of Hebei Province (A2016002025), Characteristic innovation project of Guangdong Province (2016KTSCX154), The authors also would like to acknowledge the financial support of Natural Science Foundation of Guangdong Province (2014A030311045) and project of Education Department of Guangdong Province.

References

[1] Xinbao Yu, Shaobo Li. Based on the strength Pareto evolutionary algorithm with constraints in parallel hybrid multi-objective optimization. Journal of computer applications, 2011, 31 (11): 3091-3093.

[2] Qingnian Wang, Xiaohua Zeng, Weihua Wang. The optimal mathematical modeling and simulation of energy consumption of hybrid vehicle [J]. Journal of system simulation, 2007, 19 (18): 4309-4311.

[3] Yantao Liu. Hybrid power distribution system modeling, control, and simulation study [D]. Dalian Maritime University, 2008.

[4] Tao Wang. Neural network PID control theory and its dynamic simulation research [D]. Wuhan University, 2004

[5] Jiaming Wang. Parallel ISG hybrid powertrain design and performance optimization research [D]. Shanghai Jiaotong University, 2008.

[6] Yiren Tang, Sen Wu. The classification of the series hybrid electric vehicle energy equipment and control algorithm research [D]. Bus technology, 2005, 26 (2): 4 -8.

[7] Guanci Yang, Shaobo Li. Based on the principle of Pareto optimal hybrid multi-objective optimization. Journal of Shanghai Jiaotong University, 2012 (8): 1297-1303.

[8] Xi Zhang. Vehicle Power Management; Modelling, Control and Optimization [M].China Machine Press.2013.2:239-240.

Design of Balanced Mechanisms Based on Reconfiguration for Space Applications

Dan Zhang and Bin Wei

Department of Mechanical Engineering, York University
4700 Keele Street, Toronto, Ontario, M3J 1P3, Canada

Corresponding author's e-mail: dzhang99@yorku.ca

Abstract. Often, one achieves the dynamic balancing condition by resorting to counter-devices approach, however, by doing this, one adds extra weight and therefore the inertia are increased inside the whole system, which is not cost-effective when the system is sent into space and later used in space. In this study, it is suggested one is able to achieve the reactionless condition through combining the self-balanced system. For example, the dynamic balancing condition can be realized via the reconfiguration concept. Extra counter-mass is not employed but through reconfiguring the whole structure, in this way, the system will not get to be heavy and therefore, reduce the energy costs and make the system more applicable and flexible for space applications. Based on this concept, first and foremost, one needs to balance a single component through the reconfiguration approach (i.e. decomposition process) and after that integrate the above balanced components to build the entire system (i.e. integration process). Finally, with the mechanical reconfiguration, the control laws governing the operation of the mechanism also need to be changed, so as to make whole systems more flexible when they are used in space.

1. Introduction

When mechanisms move, because the center of mass (CoM) of the system also moves and the angular momentum is not constant, vibration is therefore exist, which greatly affect the accuracy performance when the system is employed in space, especially for those are conducting high speed tasks in space applications. How to reduce those vibrations and therefore to have a flexible system used in space has become a common desire. The space flexible system concept has been widely advocated by many industries and scholars. In the space applications and its arena, the space flexible system can be simple put as this way, increase productivity and flexibility without comprising other conditions. Take the delta robot as an example, when the robot is used in space or operates in other applications, because the whole system is not dynamic balanced, the whole system will vibrate and therefore will significant affect the accuracy performance and sometimes can delay the productions.

1.1. Approach 1 - Prior to Structural Synthesis

A Dutch scholar thoroughly investigated this approach in his work [1]. Other approaches can be referred to [2-4].

1.2 Approach 2 - After Structural Design

In [5], the CRCM-based approaches were put forward. The first is the low inertia pattern approach, the second is the CRCM approach and the last is the ground CRCM balancing principle. The merit of the first approach is that the system has low inertia. The merit of the second approach is that only a single

CRCM is essential for the moment balancing. The merit of the third approach is that the system has compact construction. The single CRCM structure is not a wise approach due to a pair of gears being added to the top movable linkage instead of the ground. However, the downside of the CRCM that is close to the base structure can be concluded as follows: the CRCM gears to moment balance the top movable linkage is on the ground linkage, which causes the mechanism to be heavier. Other approaches can be referred to [6-11].

2. Dynamic balancing via reconfiguration concept

Two masses are put in the bottom, as shown in Fig. 1, and the whole system is force balanced. When we spin this system, it is also dynamically balanced. When we move one mass slightly up, and we spin the system again, it will not be dynamically balanced anymore. So how can one rearrange this system to regain the dynamic balanced condition?

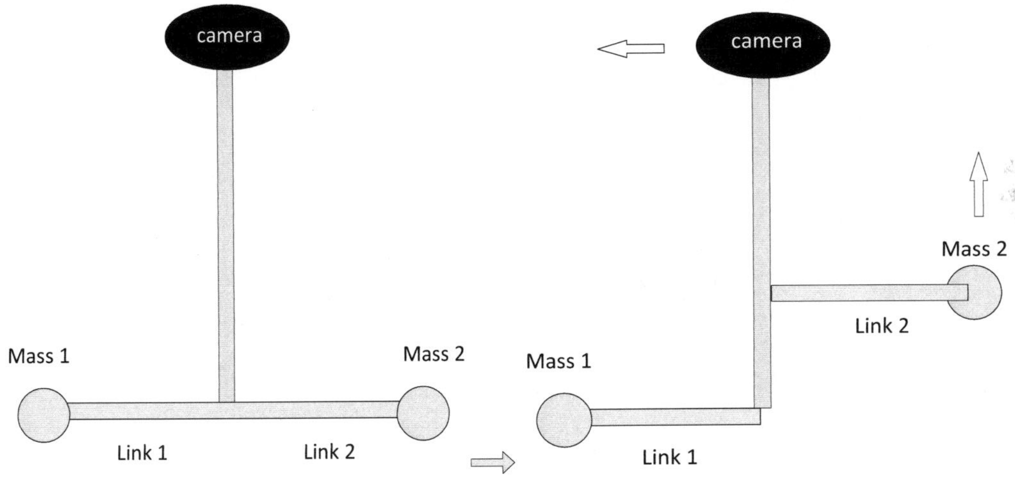

Figure 1. Simplified version of SteadiCam

Consider that one shifts the linkage 2 to a different position all the way to the top, in order to regain the balancing condition, the only way we can do is that to move the camera in a same fashion direction all the way to the left, same goes to the mass 1 but move it to the bottom. In the end, we have the same situation, it is just that there are two masses in the top and one mass in the bottom instead of two masses in the bottom and one mass in the top. From the above, it is observed that when we have the same mass relations, we can keep the system balanced.

Inspired by the above, dynamic balancing via reconfiguration is therefore put forward, which will not increase the weight of the whole system, i.e. to have a more flexible condition, so as to further improve its accuracy performance.

The purpose of using counter-weight is to make the CoM move to a fixed position. Now is it possible to achieve this by using another approach? We can re-position the linkage to make its CoM to move to a fixed position, for example, to make it move to the joint position. In this way, one does not use any extra counter-devices, and therefore the system will not get to be heavy. Based on this concept, first and foremost, one needs to balance a single component through the reconfiguration approach (i.e. decomposition process) and after that integrate the above balanced components to build the entire system (i.e. integration process).

By applying this concept to multi-DOF serial manipulator systems and other kinds of systems, new dynamic balanced multi-DOF and other systems are able to be generated, which will be presented in the following section. It is observed that the above approach will not make the system heavier or make the system have more inertia. Whereas by using the old adding counter-devices approach, one potentially puts a negative effect onto the system in terms of mass and inertia.

3. Decomposition and integration for space flexible system

Regarding the crank-slider system, it can be considered as three-linkage open chain, the end linkage is a slider which does not rotate and it only translates. Due to the fact that the end linkage does not rotate, the CoM of this linkage is in any point on this linkage. Using similar approach, one can achieve the balancing condition through reconfiguration. The crank-slider that is balanced via reconfiguration can be employed to be functioned as a Scott-Russell system to replace the traditional Scott-Russell system [8], and we integrate it into the linkage of the 3-RPR robotic mechanism to re-synthesize the whole system, as illustrated in Figure 2, which can be potentially used in space.

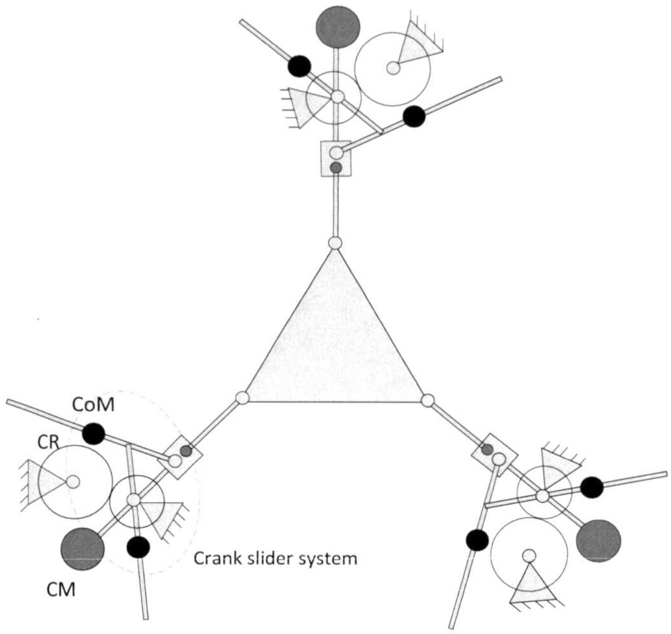

Figure 2. Dynamic balanced 3RPR planar parallel manipulator

4. Reconfiguration Control

In [12], a model reference adaptive fuzzy control system was designed for the multi-input and multi-output plant model to control an uncertain flexible joint manipulator, and the control system was approved stable by resorting to the Lyapunov approach. The novelty of the above study is that by introducing the fuzzy modelling and fuzzy model based control system into the MRAC, the new control system design can achieve the trajectory tracking effectively in spite of parameter perturbation and manipulator model uncertainty. In [13, 14], a new type adaptive control system was developed by combining an intelligent supervisory loop and the MRAC, and based on this, four different control strategies were derived, and it was applied to a single link flexible robotic manipulator. The feasibility of the design approach was verified. In [15], the control issue for an uncertain robotic manipulator with input saturations and unknown input scaling was studied by resorting to the MRAC. The MRAC was used to handle the input saturations, the unknown input scaling was rejected by the non-regressor procedure, and the robotic manipulator's uncertain dynamics were managed by the known function regressor. In [16], A MRAC algorithm using fuzzy logic adjustment of reference model parameters was designed and used in robotic manipulators. The algorithm was designed in a way in order to deal with complex environment conditions that the robotic manipulator could possibly go through. The intention of the designed control system was to ensure automatic change of the controller parameters so that they match with the current robotic manipulator environment.

There are many other types and forms of model reference adaptive control systems that are in the same family category of the above MRAC. The main strategy inside the MRAC system is that the dynamic parameters of the plant model (e.g. a space robot) are constantly being adapted and adjusted by the adaptation algorithm, and this adaptation algorithm is mainly formed by the output of the plant model and the reference model. The main purpose of adaptation is to compensate the parameters change effect in the plant due to outside disturbances. The main difficulty that one faces during the reconfiguration or adaptive control system design is the development of the adaptation algorithm. There are mainly two approaches to design the adaptation algorithm, one is based on the Lyapunov stability theorem, and the other is based on the hyper-stability theorem. In space applications, accuracy is critical for a task. For pure mechanical systems, how to have a control system that has strong adjustability and reconfigurable capability to cope with the unexpected disturbances in space is the common desire. Control reconfiguration mainly refers to adjust/change the control system structure or add an extra control component inside the whole control system in a way that the control system can handle the mechanical change. There are many articles that are dealing with hybrid control design by combining the adaptive control and other control systems, the purpose of which is to make the control system have more adaptability and move the robotic system move more smoothly.

The control system's stability performance is crucial because it is hard to develop a stable adaptation rule. Usually, for a given plant model, we can develop a stable adaptation rule, but when the plant model slight changes, for example, to cope with outside disturbances in space, the stable condition can be easily broken. This issue is worth further to be further addressed in the future, and to develop a general stable adaptation algorithm for a general type of system that are used in space and other areas. The other issue one faces is that MRAC system depends on cancelling the non-linear segments via the reference model.

5. Conclusion

In this paper, the authors presented the methodology of dynamic balancing via reconfiguration for space flexible systems, and therefore the robotic system is not getting to be heavy, which is very important when the system is sent into space and therefore, reduce the energy costs and also to make the system more flexible when used in space. After dynamically balancing a component via the reconfiguration approach, one then integrates the balanced components to build the entire system. New reactionless mechanisms can be obtained via the methodology of decomposition and integration and via resorting to reactionless structures. The mechanism and control system interconnect with one another. When one conducts the mechanism reconfiguration, the corresponding controller needs to be adapted as well. The reconfigured system is briefly discussed. Future research will focus on new model reference adaptive control system design for space flexible mechanisms and also connect the mechanism system safety issue with new model reference adaptive control, i.e. through controlling to achieve system safety when they are used in space.

Acknowledgments

The authors would like to thank the financial support from the Natural Sciences and Engineering Research Council of Canada (NSERC). The authors gratefully acknowledge the financial support from the Canada Research Chairs (CRC) program.

Reference

[1] V. Wijk, 2014, "Methodology for analysis and synthesis of inherently force and moment-balanced mechanisms - theory and applications" Ph.D. thesis, University of Twente, Netherlands.
[2] V. Wijk, S. Krut, F. Pierrot, J. L. Herder, 2013, "Design and experimental evaluation of a dynamically balanced redundant planar 4-RRR parallel manipulator," The International Journal of Robotics Research, 32, pp. 744–759.
[3] Y.N. Wu. 2003, "Synthesis and analysis of reactionless spatial parallel mechanisms," Ph.D. thesis, Laval University, Canada.
[4] C.M. Gosselin, F. Vollmer, G. Cote, Y.N. Wu, 2004, "Synthesis and design of reactionless three-degree of freedom parallel mechanisms," IEEE Transactions on Robotics and Automation, Vol. 20, No. 2, pp. 191-199.

[5] V. Wijk, J. L. Herder, 2009, "Synthesis of dynamically balanced mechanisms by using counter-rotary counter-mass balanced double pendula," ASME Journal of Mechanical Design, Vol. 131, No. 11, pp. 111003-1-8.

[6] V. Wijk, B. Demeulenaere, J. L. Herder, 2009, "Comparison of various dynamic balancing principles regarding additional mass and additional inertia," ASME Journal of Mechanisms and Robotics, Vol. 1, No. 4, pp. 041006-1 -9.

[7] V. Wijk, J.L. Herder, 2008, "Dynamic balancing of mechanisms by using an actively driven counter-rotary counter-mass for low mass and low inertia," Proceedings of the Second International Workshop on Fundamental Issues and Future Research Directions for Parallel Mechanisms and Manipulators, France, pp. 241-251.

[8] Sébastien Briot, I.A. Bonev, C.M. Gosselin, V. Arakelian, 2009, "Complete shaking force and shaking moment balancing of planar parallel manipulators with prismatic pairs," Proceedings of the Institution of Mechanical Engineers-Part K, Vol. 223, Issue 1, pp. 43-52.

[9] V. Wijk, J. Herder, 2010, "Dynamic Balancing of a single crank-double slider mechanism with symmetrically moving couplers," New Trends in Mechanism Science, Mechanisms and Machine Science, Vol. 5, pp 413-420.

[10] V. Wijk, J. Herder, 2010, "Active dynamic balancing unit for controlled shaking force and shaking moment balancing," ASME 2010 International Design Engineering Technical Conferences and Computers and Information in Engineering Conference, Canada, pp. 1515-1522.

[11] K. Wang, M. Luo, T. Mei, 2013, "Dynamics analysis of a three-DOF planar serial-parallel mechanism for active dynamic balancing with respect to a given trajectory" International Journal of Advanced Robotic Systems, 10:23. doi: 10.5772/54201, pp.1-10.

[12] C. Park, Y. Cho. "Adaptive tracking control of flexible joint manipulator based on fuzzy model reference approach". IEE Proc., Control Theory Appl., 150, (2), pp. 198–204, 2004.

[13] S. Kamalasadan. "A new generation of adaptive control: an intelligent supervisory loop approach". PhD thesis, The University of Toledo, 2004.

[14] S. Kamalasadan, A. Ghandakly. "A Neural Network Based Intelligent Model Reference Adaptive Controller". CIMSA 2004 - IEEE International Conference on Computational Intelligence for Measurement System and Applications Boston, USA, 174-179, 2004.

[15] T. Tran, S. Ge, W. He. "Adaptive control for an uncertain robotic manipulator with input saturations". Control Theory and Technology, Vol. 14, Issue 2, pp 113–121, 2016.

[16] K. Al-Olimat, A. Ghandakly. "Multiple model reference adaptive control algorithm using on-line fuzzy logic adjustment and its application to robotic manipulators". 37th IAS Annual Meeting. Conference Record of the Industry Applications Conference, 1463-1466, 2002.

IOP Conf. Series: Materials Science and Engineering **235** (2017) 012015 doi:10.1088/1757-899X/235/1/012015

Research on Acceleration Compensation Strategy of Electric Vehicle Based on Fuzzy Control Theory

Tianjun Zhu[1,3], Bin Li[2], Changfu Zong[1] and Zhicheng Wei[3]

[1] Department of Electronic Information and Electrical Engineering, ZhaoQing University, ZhaoQing, 526061, China
[2]CONCAVE Research Center, Department of Mechanical and Industrial Engineering, Concordia University, Montreal, Quebec, H3G 1M8, Canada
[3]College of Mechanical and Equipment Engineering, Hebei University of Engineering, Handan, 056038, China

Corresponding e-mail: happy.adam2012@hotmail.com

Abstract. Nowadays, the driving technology of electric vehicle is developing rapidly. There are many kinds of methods in driving performance control technology. The paper studies the acceleration performance of electric vehicle. Under the premise of energy management, an acceleration power compensation method by fuzzy control theory based on driver intention recognition is proposed, which can meet the driver's subjective feelings better. It avoids the problem that the pedal opening and power output are single correspondence when the traditional vehicle accelerates. Through the simulation test, this method can significantly improve the performance of acceleration and output torque smoothly in non-emergency acceleration to ensure vehicle comfortable and stable.

1. Introduction

The issue of environmental pollution is currently more and more serious. Under the pressure of environmental problems, more and more people pay attention to the electric vehicles. Various automotive manufacturers are increasing investment in pure electric vehicle. Battery electric vehicle research has become an important direction. Acceleration performance is an important indicator for car evaluating. On the one hand, people hope that vehicles can accelerate smoothly based on people's intention for safety and comfort, on the other hand, people hope that vehicles can speed up in the power demand. Therefore, the judgment by controller is very important to the driver's intention recognition and power output.

This paper presents a method for solving different dynamic response problems of vehicles under different driver intentions. The driver's intent is reflected in the opening of the accelerator pedal. The controller controls the power output through the pedal opening MAP. At the same time, the change rate of the accelerator pedal is judged, and the compensation of the vehicle power is determined by the reasoning of human experience, which Increase the output of power in acceleration to meet the driver's needs.

2. The acceleration compensation strategy

The driver's intention to the car is reflected in the size of the pedal opening and the pedal opening rate, which is very complex. Traditional control methods have limitations that the acceleration pedal and

IOP Conf. Series: Materials Science and Engineering **235** (2017) 012015 doi:10.1088/1757-899X/235/1/012015

power output have only a single mapping relationship. This paper takes the electric platform of Hebei University of Engineering as the test object, combined with acceleration pedal opening, acceleration pedal opening rate and battery SOC value to research. When the value of SOC is low, it appropriate to reduce the compensation or no compensation to ensure that the vehicle run farther. When the value of SOC is high, the compensation value can be appropriately increased to ensure the power performance. Therefore, with a variety of input variables, the application of the fuzzy control theory can better reflect the control intention than the classical control theory. It is the first choice of nonlinear control problem without relying on mathematical model.

3. Fuzzy Controller Design

Fuzzy controller is the core of fuzzy reasoning,whichstructure is shown in Figure 1. The input parametersare pedal change rate, SOC of battery and the accelerator pedal opening degree. The input of fuzzy controller is normalized. The rate of change of acceleration pedal normalized to K, which is equal to dθ/dt. K1and K2 are the upper and lower threshold of the change rate.When the pedal change rate is lower than K1, it is equal to K1, when the pedal change rate is higher than K2, it is equal to K2.The domain of k is [0, 1]. 0 indicates that the pedal change rate is below or equal to k1,1 indicating that the pedal change rate is above or equal to K2. The input of SOC is the battery energy state after normalization, which domain is [0,1]. 0 indicates that the SOC value is less than or equal to 20%, and 1 represents the saturation state of the battery. The input of θ is the accelerator pedal opening after normalization, which domain is [0,1]. 0 indicates that the pedal opening is at position of 0% and the 1 represents the maximum opening position of the pedal. K is the proportional parameter after fuzzy inference, which domain is [0,1]. K is the proportional factor of compensating power. 0 represents no compensation and1 represents the maximum compensation.

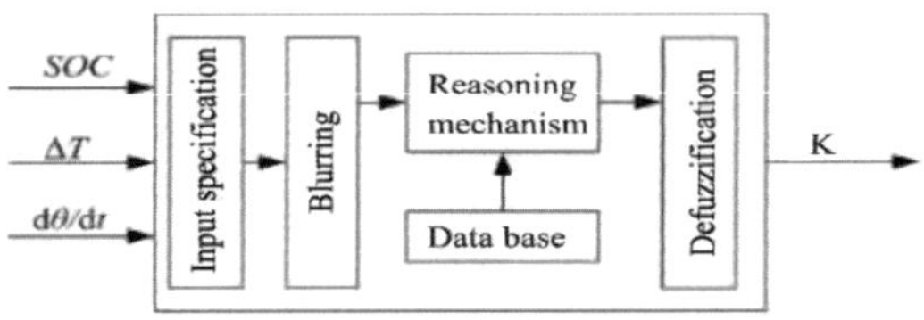

Figure 1. Fuzzy controller structure

The selection of membership function has many methods, such as two element comparison sort method, assignment method, fuzzy statistics method, etc [1-2]. It is influenced not only by objective rules but also by subjective factors of designers. Based on the reference of other literatures and the similar control cases [3-4], the Gauss type function is chosen to be applied to input variables and output variables. The value of SOC, pedal opening and pedal change rate have the same Gauss points [5]. Li represents small, Ml represents medium little, Me represents medium, Mb represents medium big, Hi represents high. The input and output fuzzy subsets are as follows:

SOC:{Li,Me,Hi}

k: {Li,Me,Hi}

θ: {Li,Me,Hi}

K : {Li,Ml,Me,Mb,Hi}

The input and output membership functions are shown in Figure 2-3.

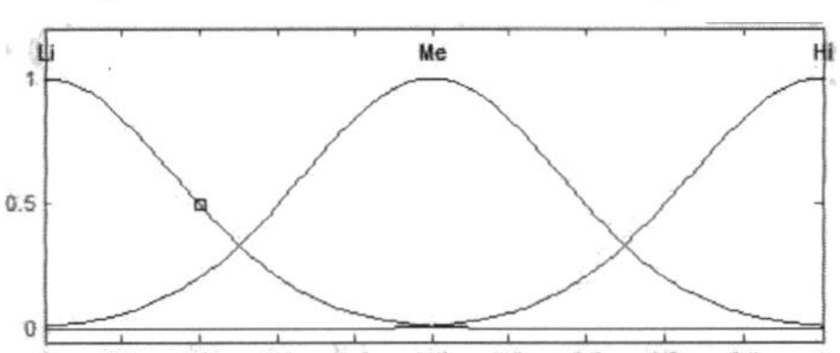

Figure 2. Input membership function

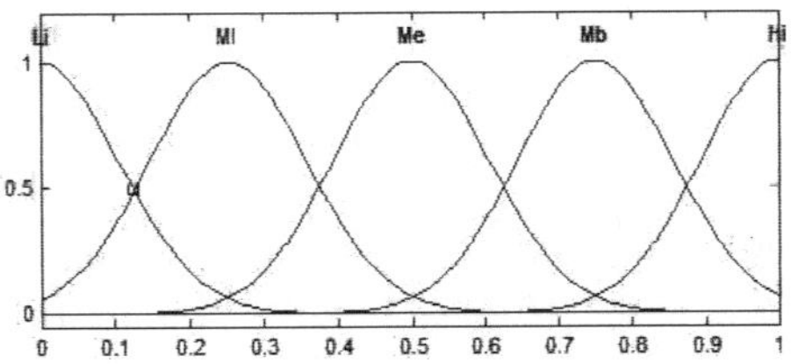

Figure 3. Output membership function

The conditions for the accelerator pedal compensation are described as follows:

(1)When the value of SOC is low and the pedal opening conversion is not obvious, there is no acceleration power compensation. When the accelerator pedal opening change is large, which means that the driver has a sharp acceleration intention, there is a small amount of compensation appropriately.

(2) When the battery SOC value is high or medium, the compensation focuses on the acceleration pedal changes. When the pedal change is not obvious, there is a small amount of compensation. When the pedal opening and pedal change rate are large, it is considered that the intention of acceleration is obvious, so the power compensation is high.

Table for fuzzy logic rules is shown as Table 1. Fuzzy controller reasoning surface graph is shown as Figure 4.

k	θ	K	k	θ	K	k	θ	K
SOC=Li			SOC=Me			SOC=Hi		
Li	Li	Li	Li	Li	Li	Li	Li	Li
Li	Me	Li	Li	Me	Li	Li	Me	Ml
Li	Hi	Li	Li	Hi	Ml	Li	Hi	Me
Me	Li	Li	Me	Li	Ml	Me	Li	Ml
Me	Me	Ml	Me	Me	Me	Me	Me	Me
Me	Hi	Ml	Me	Hi	Mb	Me	Hi	Mb
Hi	Li	Li	Hi	Li	Ml	Hi	Li	Ml
Hi	Me	Ml	Hi	Me	Me	Hi	Me	Me
Hi	Hi	Me	Hi	Hi	Hi	Hi	Hi	Hi

Table 1. The optimization target and weight coefficient

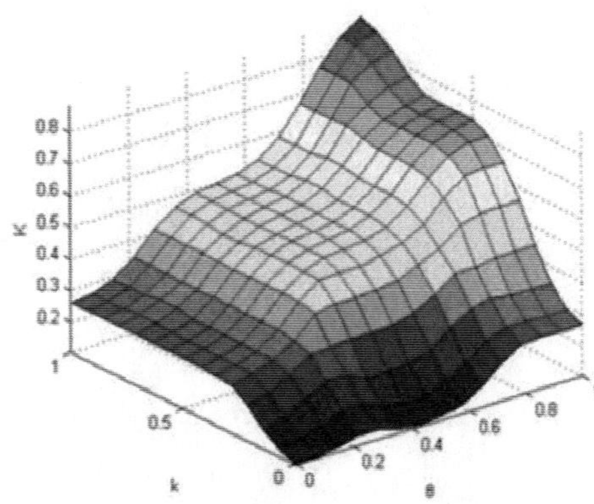

Figure 4. Fuzzy inference surface

4. Control strategy

The output of power is embodied in the request of the vehicle management System (VMS) for the output torque of the motor controller [6]. The acceleration compensation means that increase an appropriate torque value based on the torque output of the motor, which make the motor speed of the vehicle to the target speed fleetly. This paper uses the method of control input power to motor to change the torque. Electric vehicle power flow is shown in Figure 5.

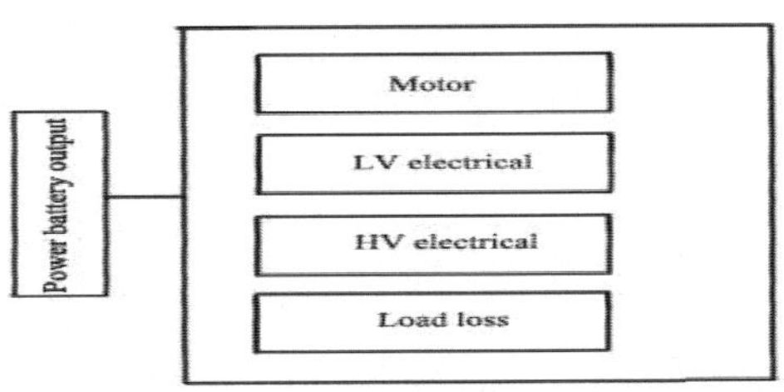

Figure 5. Vehicle power flow

The battery management system controls the output of the power according to the command of the vehicle controller. P_{motor} represents the power consumed by the motor. The power consumed by the low voltage load and the high voltage load shall not be distinguished in this article, which recorded as P_e. P_w represents the loss of energy in transmission. The power request of the acceleration compensation system based on fuzzy controller is shown in Figure 6. In the absence of compensation, the electric vehicle output the standard request power through the pedal MAP, when the fuzzy controller is involved in the calculation of the parameters K, the electric vehicle compensation power is KP_{con}. P_{con} is the smallest unit of power compensation. In order to better reflect the original intention of energy management, P is set to dynamic variables through the SOC-MAP diagram calculation. P_{con} is a function mapping of the battery SOC value and the same size as it. When the SOC value of the battery is large, the P_{con} value is the largest. When the SOC value is less than 25%, the Pcon value is the smallest. Therefore, the request for compensation for power:

$$P_{motor1} = P_{motor} + KP_{con} \tag{1}$$

P_0 represents the actual output power of the motor, which is calculated according to the feedback of the motor speed n and the current torque T:

$$P_0 = nT / 9550 \tag{2}$$

The actual demand of the vehicle is:

$$P_{re} = P_{motor1} + P_e + P_w - P_0 \tag{3}$$

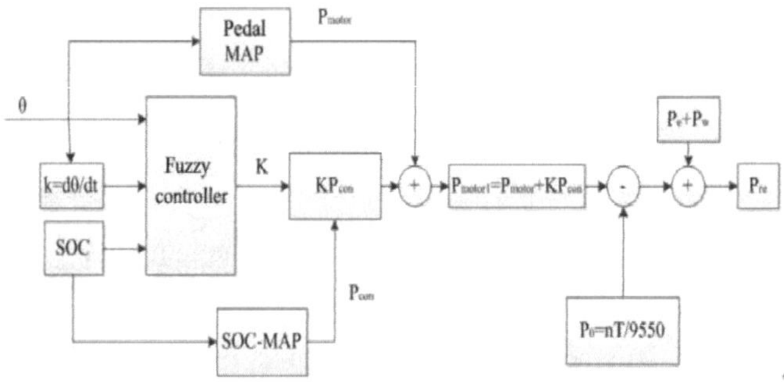

Figure 6. Power request structure based on fuzzy control

5. Simulation results

The simulation model of acceleration compensation of electric vehicle is built by Matlab/Simulink based on the structure of fuzzy control. Pedal input is simulated by the Ramp module in Simulink as shown in Figure 7. The pedal opening is 90% and the slope is the rate of change of the accelerator pedal.

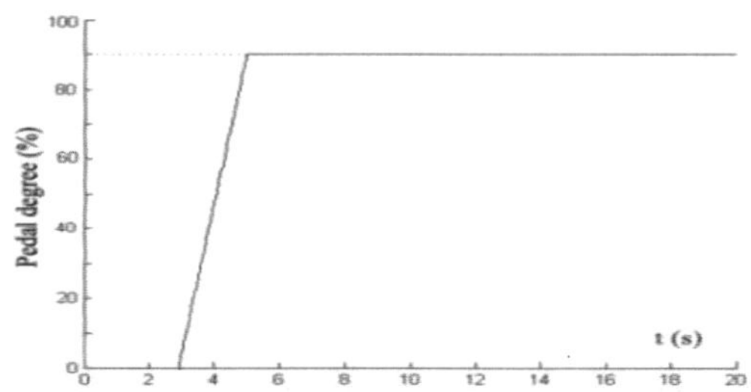

Figure 7. Accelerator pedal simulation

The simulation results are shown in Figure 8-9.

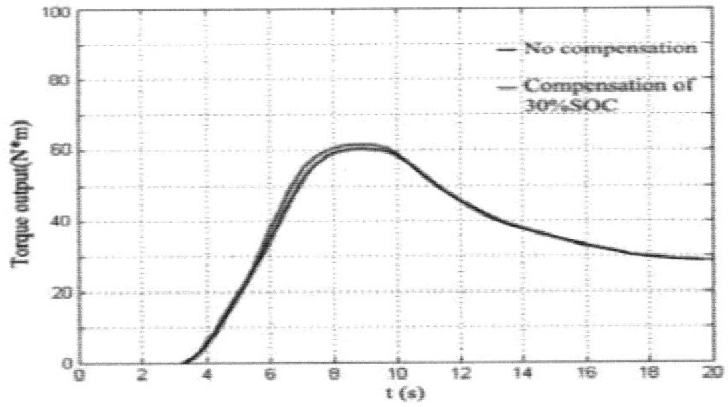

Figure 8. Torque output of 30% SOC

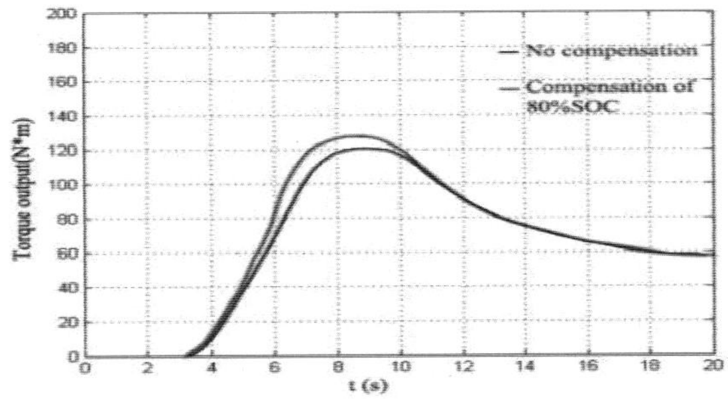

Figure 9. Torque output of 80%SOC

As can be seen from the Figure 8-9, Acceleration compensation is still a small value even if the pedal is open to 90% when the SOC value is relatively small. A small increase in power compared to the non- compensatory control strategy. When the SOC value is higher, the compensation value of torque increases and the output power increases obviously.

6. Conclusions

On the basis of traditional electric vehicle control theory, in this paper, the acceleration change rate is judged and the battery state is used to realize the compensation of driver's intention recognition for power output. The corresponding simulation model is established based on Simulink. A fuzzy controller is designed in this paper, which is used to realize the dynamic control of multiple nonlinear constraints. The test results show that the control strategy can compensate the torque in the acceleration and can be adjusted dynamically according to the driver's intention. In the application, according to the different types of vehicles and the driver's needs, adjust the SOC-MAP to realize the change of the Pcon, which make the power compensation adjustment more convenient. The model is valuable to the research and application.

Acknowledgments

The authors would like to acknowledge the financial support of Collaborative Innovation and Platform Environment Construction Project of Guangdong Province (2015A050502053) , Hebei science and technology plan project(17394501D), Science and technology research project of Hebei higher education (ZD2017213) Hebei provincial High Level Talents Foundation of Hebei Province (A2016002025), Characteristic innovation project of Guangdong Province (2016KTSCX154), The authors also would like to acknowledge the financial support of Natural Science Foundation of Guangdong Province (2014A030311045) and project of Education Department of Guangdong Province.

IOP Conf. Series: Materials Science and Engineering **235** (2017) 012015 doi:10.1088/1757-899X/235/1/012015

References

[1] Dong Bing,Tian Yan-tao,Zhou Chang-jiu.Torque correcting strategy in driving control of new energy vehicle. 2013 Fifth Conference on Measuring Technology and Mechatronics Automation . 2013

[2] Cao, Jian,Emadi, Ali.A new battery/ultracapacitor hybrid energy storage system for electric, hybrid, and plug-in hybrid electric vehicles. IEEE Transactions on Power Electronics . 2012

[3] Banvait H, Anwar S,Chen Yaobin.A rule-based energy management strategy for plug-in hybrid electric vehicle (PHEV). American Journal of Infection Control . 2009

[4] Rosenberger M,et al.Combining Regenerative Braking and Anti-Lock Braking for Enhanced Braking Performance and Efficiency. SAE Technical Paper 2012 -01-0234 . 2012

[5] Xiaoyu Huang,Junmin Wang. Model predictive regenerative braking control for lightweight electric vehicles with in-wheel motors[J]. Proceedings of the Institution of Mechanical Engineers, Part D: Journal of Automobile Engineering . 2012 (9).

[6] Saeid Bashash,Scott J. Moura,Joel C. Forman,Hosam K. Fathy. Plug-in hybrid electric vehicle charge pattern optimization for energy cost and battery longevity[J]. Journal of Power Sources . 2010.

IOP Conf. Series: Materials Science and Engineering **235** (2017) 012016 doi:10.1088/1757-899X/235/1/012016

Analysis of Interval Control System Robust Quality Indices on the Base of Root and Coefficient Methods

S A Gayvoronskiy[1], T Ezangina[1] and I Khozhaev[1]

[1]Department of Automatics and Computer Systems, National Research Tomsk Polytechnic University, Lenin Avenue 30, Tomsk, Russia

Corresponding email: eza-tanya@yandex.ru

Abstract. Considered paper is dedicated to interval control system quality analysis with the help of root locus method and method of coefficient estimates. On the base of the first method, an algorithm of determining examinable vertices of interval characteristic polynomial coefficients polytope, which images determine robust root quality indices. On the base of coefficients method, sufficient conditions, linking coefficients of interval characteristic polynomial coefficients with quantitative estimates of robust root quality indices. With the help of both developed approaches, a robust control quality of unmanned underwater vehicle control systems was analyzed.

1. Introduction

Control systems must sustain their manipulated variables despite interval parametric uncertainty. Consequently, a problem of analyzing robust root quality indices of an interval control system (ICS) is highly relevant. Robust quality indices can be clearly analyzed with the help of root approach, which is based on an examination of ICP poles, migrating inside their allocation areas of complex plane (figure 1). There are some methods of robust quality indices (the degree of robust stability α and the degree of robust oscillation $\mu = tg\varphi$) analysis, which are based on an interval characteristic polynomial (ICP) roots examining [1-10].

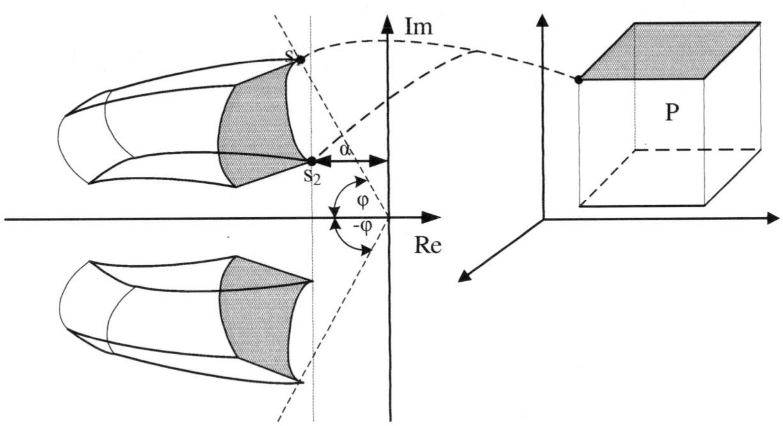

Figure 1. System poles allocation areas.

IOP Conf. Series: Materials Science and Engineering **235** (2017) 012016 doi:10.1088/1757-899X/235/1/012016

A problem of robust root quality analysis for an ICS with interval uncertainty of ICP coefficients can be solved by analyzing vertices of polytope P 10. Let us notice, that examining of all of 2^m vertices of parametric polytope of ICP, having m interval parameters, can be intensive and time-consuming, so, it is advisable to use methods of examinable vertices number reduction [1-4; 9-10].

Another method of robust root quality indices analysis is to calculate them with the help of quantitative estimators. Such estimations are proposed to be performed with the help of coefficient method [11]. This method is based on solving a system of algebraic inequalities, containing constant characteristic polynomial coefficients. To apply this method to an ICS analysis, it must be extended to characteristic polynomials with interval coefficients.

Considering this, a problem of examining an accuracy of previously mentioned methods of ICS robust quality indices analysis is highly relevant.

2. Calculating robust root quality indices on the base of examinable vertices of interval parameters polytope

Let us assume, that an ICP is written as follows:

$$D(s) = d_n s^n + d_{n-1} s^{n-1} + \cdots + d_0, \tag{1}$$

where $d_i \in [d_{i\min}, d_{i\max}]$, $i \in \overline{0,n}$ and are included in rectangular polytope P_m with 2^m vertices $(m=n+1)$. Let us consider a transfer function of an edge of P_m

$$W_i^{V_q}(\Delta d_i, s) = \frac{\Delta d_i s^i}{D^{V_q}(s)}, \tag{2}$$

where is an increment if an i-th interval coefficient; $D^q(s) = \sum_i d_i^q s^i + \sum_p d_p s^p$, $p \in \overline{0,n}$, $p \neq i$ – vertex polynomial, d_p – constant coefficients of (1), d_i^q - values of interval coefficients in a V_q vertex.

Let us propose a procedure of determining an edge route of a parametric polytope, connecting examinable vertexes. This procedure consists in performing following steps.

1) Calculating a complex root of (1), corresponding to root node U_q, $q \in \overline{1,2^m}$ q – number of a vertex, for a random vertex V_q.
2) Determining m exit angles of edge branches Θ_i^{Vq}, $i \in \overline{0,n}$ for all ICP coefficients on the base of (2) for U_q. These angles can be determined with the help of equations from 10.
3) Checking conditions

$$\Theta_{\max}^{V_q} - \Theta_{\min}^{V_q} < 180°, \tag{3}$$

where $\Theta_{\max}^{V_q}, \Theta_{\min}^{V_q}$ are minimal and maximal exit angles. If the condition (3) is satisfied, then U_q is included in border of (1) roots allocation areas S_r. If any of (3) conditions is not satisfied, then another vertex P_m should be chosen, and steps 1 – 3 must be repeated.

4) Composing an ascending sequence of exit angles of edge branches Θ_{di}^{Vq} of root node U_q and according set of interval coefficients.
5) Finding bounds G_{ij}, which R_i^q and R_j^q edge images may intersect.

In 10 it is stated, that edge branches R_i^q and R_j^q intersect in nodes U^* only inside those bounds G_{ij}, which are created with ICP coefficients, which satisfy a following condition:

$$|i - j| \geq 3. \tag{4}$$

6) Composing an edge route P_m, according to the sequence of interval coefficient from step 5 and considering conditions (4).
7) Determining a set of vertexes V_q, which are included in an edge route.
8) Plotting an image of V_q vertexes on a complex plane and estimating robust root quality indices of the ICS.

3. Estimating robust root quality indices on the base of interval coefficients of characteristic polynomial

Let us consider interval coefficient estimators of root quality indices: stability indices $[\lambda_i]$, $i \in \overline{1, n-2}$ and oscillability indices $[\delta_p]$, $p \in \overline{1, n-1}$. Stability indices can be calculated with the help of formula: $[\lambda_i] = [d_{i-1}][d_{i+2}] / [d_i][d_{i+1}]$, $i \in \overline{1, n-2}$. Interval oscillability indices can be calculated as follows: $[\delta_p] = [d_p^2] / [d_{p-1}][d_{p+1}]$, $p \in \overline{1, n-1}$.

It is proposed to calculate a low estimate of a considered ICS robust stability degree on the base of $[\lambda_i]$. To do this, in [12] a following statement was proved.

Statement 1. A low estimate of robust stability degree η^* can be calculated by solving a minimax problem $\eta^* = \min_i \max \eta_i, i \in \overline{1, n-2}$ for a following system of inequalities

$$\begin{cases} \overline{\lambda^{**}{}_i(\eta_i)} = \dfrac{\overline{d_{i-1}}\,\overline{d_{i+2}}}{\left(\underline{d_i} - \overline{d_{i+1}}(n-i-1)\eta_i\right)\left(\overline{d_{i+1}} - \overline{d_{i+2}}(n-i-2)\eta_i\right)} = \lambda^*, \ i \in \overline{1, n-2}, \ \lambda^* \approx 0.465; \\ \underline{F_g(\eta_i)} = \underline{d_g} - \overline{d_{g+1}}(n-g-1)\eta_i \geq 0, g \in \overline{1, n-1}; \\ \underline{F_0(\eta_i)} = \underline{d_0} - \overline{d_1}\eta_i + 2\underline{d_2}\dfrac{\eta_i^2}{3} \geq 0. \end{cases} \quad (5)$$

On the base of sufficient condition of acceptable oscillability degree, a statement, allowing to calculate a high estimate of robust oscillability degree (1), is derived in [12].

Statement 2. A high estimate of robust stability degree δ^* can be calculated as $\delta^* = \min_p \delta_p$, $p \in \overline{1, n-1}$ on the base of system solution

$$\underline{d_p^2} - \delta_p \overline{d_{p-1}}\,\overline{d_{p+1}} = 0, \ p = \overline{1, n-1}. \quad (6)$$

Root allocation angle function of δ^* values is shown in the figure 2.

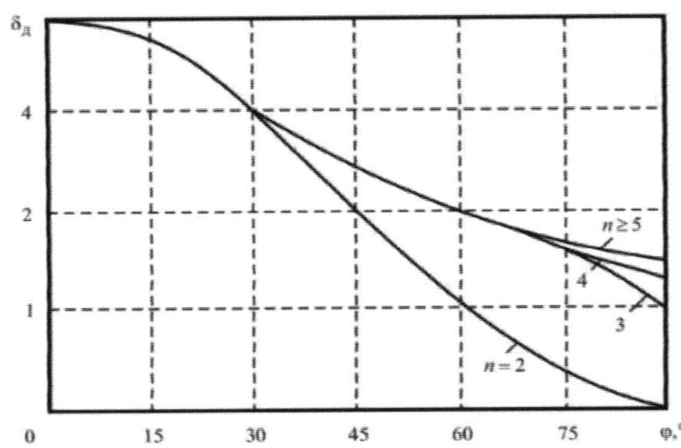

Figure 2. Root allocation angle function of δ^* values.

IOP Conf. Series: Materials Science and Engineering **235** (2017) 012016 doi:10.1088/1757-899X/235/1/012016

4. Robust control quality analysis of a unmanned underwater vehicle control system

Let us consider an unmanned underwater vehicle (UUV) vertical positioning system, which is described in [13]. Its ICP can be written as follows:

$$D(s) = [d_3]s^3 + [d_2]s^2 + [d_1]s^1 + [d_0],$$

where $[d_3] = [894;930]$; $[d_2] = [3646.41;21949]$; $[d_1] = [53444.15;297856.62]$; $[d_0] = [17120.54;94564.08]$.

Let us calculate a robust stability degree α and robust oscillability degree μ by projecting examinable vertices Pm on the complex plane. To do this, a newly developed algorithm will be applied. As a result, an edge route of the considered system will be plotted (figure 3). The limits of the set of interval coefficients in examinable vertices, shown in Figure 3, are presented in Table 1.

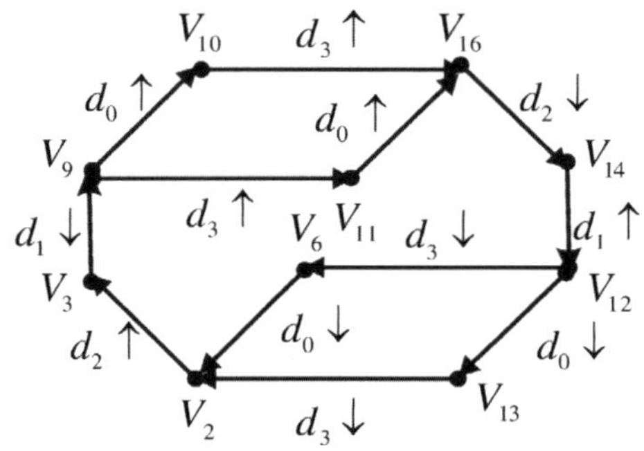

Figure 3. Edge route of the UUV vertical positioning system.

Table 1. The limits of the set of interval coefficients in of examinable vertices

# V_q	The limits of d_i, $i \in \overline{0,3}$	# V_q	The limits of d_i, $i \in \overline{0,3}$	# V_q	The limits of d_i, $i \in \overline{0,3}$
V_3	$[\underline{d_3};\overline{d_2};\overline{d_1};\underline{d_0}]$	V_{14}	$[\overline{d_3};\underline{d_2};\underline{d_1};\overline{d_0}]$	V_6	$[\underline{d_3};\underline{d_2};\overline{d_1};\overline{d_0}]$
V_9	$[\underline{d_3};\overline{d_2};\underline{d_1};\underline{d_0}]$	V_{12}	$[\overline{d_3};\underline{d_2};\overline{d_1};\overline{d_0}]$	V_{11}	$[\overline{d_3};\overline{d_2};\underline{d_1};\underline{d_0}]$
V_{10}	$[\underline{d_3};\overline{d_2};\underline{d_1};\overline{d_0}]$	V_{13}	$[\overline{d_3};\underline{d_2};\overline{d_1};\underline{d_0}]$		
V_{16}	$[\overline{d_3};\overline{d_2};\underline{d_1};\overline{d_0}]$	V_2	$[\underline{d_3};\underline{d_2};\overline{d_1};\underline{d_0}]$		

According to the figure 3, 10 vertices must be examined. Projecting them on a complex plane and examining their allocation resulted in following values of robust stability degree $\alpha = 0.057$ in V_2 and robust oscillability degree $\mu = 9.83 (\varphi = 84^\circ)$ in V_{12}.

After that, with the help of the second approach, let us calculate quantitative estimates of robust stability and oscillability degrees. Solving a (5) system resulted in a low estimate of robust stability degree $\eta^* = 0.056$. Solving (6) system resulted in a high estimate of robust oscillability degree $\delta^* = 0.048$, which corresponds with system poles allocation in a sector of $\varphi = 89^\circ$.

Comparative analysis of robust root quality indices, being calculated with the help of two methods, shown us, that accurate value of these indices and their estimates do not differ significantly for systems with interval ICP coefficients uncertainty.

5. Conclusion

There are major three conclusions that the paper figures out.

1) An algorithm of composing an edge route of an ICS parameters polytope, including all examinable vertices, which determine robust root quality indices, was developed on the base of interval expansion of root locus method.
2) Interval coefficient indices are proposed to be used for ICS robust control quality indices estimation. In order to calculate these coefficient indices, sufficient conditions, linking ICP coefficients with low estimate of robust stability degree and high estimate of robust oscillability degree, were derived. A method of estimating robust root quality indices was developed on the base of these sufficient conditions.
3) A robust control quality of UUV control systems was analyzed with the help of developed methods. It was determined, that values of ICS root quality indices, calculated with the help of two methods, do not differ significantly. However, a coefficient method of quality indices estimation is less time-consuming and complex, than interval expansion root locus method. So, it is advisable to use a coefficient method to analyze a robust control quality of ICS of high degree.

Acknowledgments
The reported study is supported by the Ministry of Education and Science of Russian Federation (project #2.3649.2017/PCh).

References

[1] Ackermann J 1980 Parameter space design of robust control systems *IEEE Transactions on Automatic Control* vol. 25 # 6 pp 1058-1072.
[2] Barmish B R and Tempo R 1990 The robust root locus *Automatica* vol. 26 # 2 pp 283-292.
[3] Rimsky G V and Nesenchuk A A 1996 Root locus methods for robust control systems quality and stability investigations *Proc. of IFAC 13th Triennial World Congress* (*San Francisco, USA*) pp 469-474.
[4] Soh Y.C. Foo Y K 1989 Generalized edge theorem *Systems and Control Letters* vol. 12 # 3 pp 219-224.
[5] Neimark Y I 1992 Robust stability and D-partition *Automation and Remote Control* vol. 53 # 7 pp 957–965.
[6] Efimov S V, Pushkarev M.I. 2011 Determining direct measures of performance based on the location of zeros and pole of the transfer function *Optoelectronics, Instrumentation and Data Processing* vol. 47 pp. 297–302.
[7] Juang B Y 2010 Robustness of pole assignment of an interval polynomial using like λ˜–Degree feedback gain based on the Kharitonov theorem *Proc of the SICE Annual Conference 2010* SICE 2010 *(Taipei; Taiwan)* pp. 3475-3484.
[8] Foo Y K, Soh Y C 2001 Robust root clustering of interval plants with first-order compensators *IEE Proceedings: Control Theory and Applications* vol. 148 issue 4 pp. 315-317.
[9] Polyak B T and Scherbakov P S 2002 *Robust stability and control* (Moscow: Science) p 303.
[10] Vadutov O S and Gayvoronskiy S A 2003 Application of edge routing to stability analysis of interval polynomials *Journal of Computer and Systems Sciences International* vol 42 # 6 pp 833-838.
[11] Petrov B N, Sokolov N I, Lipatov A V and other objects 1986 *Automatic control systems with variable parameters: Engineering methods of analysis and synthesis* (Moscow: Mechanical Engineering) p. 256.
[12] Gayvoronskiy S A, Ezangina T A 2015 The algorithm of analysis of root quality indices of high order interval systems *Proc of the 2015 27th Chinese Control and Decision Conference (CCDC 2015) (China) (New York: IEEE)* p. 3048-3052.
[13] Gayvoronskiy S A, Khozhaev I and Ezangina T Motion control system for a remotely operated vehicle with interval parameters *International Journal of Mechanical Engineering and Robotics Research* submitted for publication.

Institute of Physics
Dirac House, Temple Back
Bristol BS1 6BE UK

ISSN: 1757-8981
ISBN 978-1-5108-4778-1